Presented to 獻給

May the grace, mercy and peace of God be with you and preserve
you and your loved ones in good health always.

By 致贈人

Date 致贈日期

Occasion 祝福您

For we are God's workmanship, created in Christ Jesus to do good works,
which God prepared in advance for us to do.

KC健康廚房

從 零 開始 學做菜！

向食物借油×借水的健康美味烹調法

Love @ First Bite

新加坡原味料理達人&世界無水烹調冠軍

Kevin & Claire —— 合著

目錄 Contents

006　　【推薦序1】充滿愛的原味料理/Original Flavor Cuisine that is Full of Love ◎陳美鳳/Chen, Mei-Fung

007　　【推薦序2】認真熱情的原味料理達人/The Earnest, Sincere Fxperts of Original Flavor Cuisine ◎陳凱莉/ Kelly Chen

009　　【自　　序】不能錯過的烹飪新浪潮/ New Wave Cuisine~ KC Natural & Heavenly Fare ◎Kevin & Claire

018　　【本書使用說明】/How to use this Book

PART1

讓原味更美味的基礎營養與採購學
MENU PLAN and NUTRITION BASICS as a PRIMER on TRUE NATURAL TASTES to PERFECTION

028　　基礎營養學，先認識飲食金字塔/Nutrition Basics starts with Understanding the Food Guide Pyramid

030　　基礎採購學，先練習開菜單/
　　　　Basic Shopping starts with Creating a Menu Plan

PART2

讓原味更美味的食材處理、刀工與烹調要領
Correct Food Preparation, Cutting Instructions and Healthy Cooking Tips to Enhance the Natural-Flavor

044　　保持食物營養的原味烹調基礎/
　　　　The Foundation for Maximum Retention of Nutrients for Natural-Flavored Cooking

059　　原味烹調的簡單基礎刀工/
　　　　Simple Basic Cutting Skills for Natural-Flavored Cooking

065　　向食物借油借水的烹調原理/
　　　　The Principle of "Greaseless and Waterless Cooking"

066　可以向哪些食物借油烹調圖解示範/
　　　"Greaseless Cooking" with Photo Tutorial
069　可以向哪些食物借水烹調圖解示範/
　　　"Waterless Cooking" with Photo Tutorial
072　讓原味健康升級成五星級料理/
　　　Upgrading Natural and Nutritious, Flavor-Filled
　　　Homely Fare to 5-Star Cuisine

PART3

讓原味更豐富的調味料、辛香料與佐醬
Original Flavor Enhancing Seasonings, Spices and Dips

076　讓原味加值的調味料/Original Flavor Enriching
　　　Seasonings
080　常用的基礎調味料/Commonly Used Seasonings
092　讓原味加分的辛香料/Original Flavor Enhancing
　　　Spices and Fresh Herbs
100　讓原味加分的佐醬/Original Flavor Enriching
　　　Sauces

讓原味更美味的50道食譜

向食物借油原味料理/Greaseless cooking

110　黃薑脆皮雞腿/Crispy Turmeric Chicken
　　　Drumstick
112　香茅燉雞/Chicken Stew with Lemon Grass
114　楓燒雞翅/Maple-Barbecued Chicken Wings
116　涼拌芝麻雞絲/Sesame Chicken Salad
118　迷迭香嫩煎梅花豬排/Grilled Pork Picnic
　　　Steak with Rosemary

120	脫脂楓糖醋燒小排/Defatted Maple Vinegar Sparerib
122	和風銀芽肉絲/Shredded Pork and Sprouts
124	香椿醬烤羊小排/Roasted Lamb Rack in Cedar Pesto
126	原味牛肩肉骨茶湯/Chuck Flat Iron Beef Broth
128	椒鹽烤無骨牛小排/Grilled Peppercorn Boneless Short Ribs
130	香根牛肩胛肉/Beef Chuck Flap Stir-Fry with Chinese Parsley
132	椒麻黃魚/Szechwan Peppercorn Yellow Fish
134	香草魚排佐蘋果番茄沙沙醬/Codfish with Apple Tomato Salsa
136	清烤魩仔魚vs.醬燒丁香魚/Pan-Fry Larval Fish vs. Saute of Silver Anchovy with Peanuts
138	洋蔥煎蛋/Onion Omelet
140	鐵板蔬菜豆腐/Vegetarian Tofu Teppanyaki
142	椒鹽烤明蝦/Pan Grilled Peppery Salt King Prawns
144	奶香淡菜盅/Grilled Creamy Mussels
146	西芹番茄炒墨魚/Pan-Broiled Squid with Celery and Tomatoes Stir-Fry
148	橙汁軟絲/Lemony Orange Glazed Squid

向食物借水原味料理/Waterless Cooking

150	鍋烤龍蝦/Pan-Grilled Lobster
152	鮮味蝦/Shrimp Delight
154	南瓜薯泥沙拉/Mixed Pumpkin Potato Salad
156	苦盡甘來/Sweetened Bitter Gourd
158	滑蛋絲瓜/Loofah Gourd in Egg Smoothie
160	胡瓜蝦仁炒蛋/Big Cucumber with Shrimp Omelet
162	魚香茄子/Teppanyaki Eggplant Topping with Saucy Minced Pork

164 鍋烤栗子/Pan-Grilled Chestnuts

166 芋頭南瓜派/Taro Pumpkin Pie

168 冰鎮柴魚菠菜/Iced Spinach Salad with Bonito Flakes

170 脆炒皇宮苗/Crisp "Huang Kong" Spinach

172 芥藍炒蘑菇/Cabbage Mustard with Button-Mushroom Stir-Fry

174 彩虹菜鍋/Rainbow Vegetarian Skillet

176 蒜香木須高麗菜/Cabbage with Black Fungus & Garlic Stir-fry

178 五福臨門/Dandy Five

180 柴燒杏鮑菇/Glazed King Oyster Mushrooms with Bonito Flakes

182 素煎雙菇/Fresh Mushroom Surprise Grille

184 干炒四季豆/French Beans with Bean Curd Stir-Fry

186 秋葵塔/Towered Gumbos

188 紅豆糯米甜湯/Glutinous Red Bean Dessert

向食物借油借水原味料理/Greaseless and Waterless Cooking

190 紅酒燉牛腩/Rib Finger Stew in Red Wine

192 蘿蔔封肉/Braised Pork with Carrot and White Radish

194 彩椒蘆筍燴鮮貝/Pan-Seared Scallops with Colorful Peppers and Asparagus Saute

196 原味鮮貝盅/Simple Clam Delight

198 鮮烤黃金蟹/Grilled Canadian Golden Crab

200 新加坡海南雞/Singaporean-Style Hainanese Chicken

202 乾煎鮭魚香椿醬炒飯/Fried Cedar Garlic Rice with Pan-Broiled Salmon

204 原味什錦炒米粉/Natural-Flavored Assorted Vegetables Fried Rice-Flour Noodles

206 南瓜船/Pumpkin Boat

208 全麥無酵鬆餅/Healthy Whole Wheat Pancakes

210 【附表1】一週家庭菜單規劃表/ATTACHMENT 1: WEEKLY FAMILY MENU PLAN

212 【附表2】我的家庭一週採買清單/ATTACHMENT 2：WEEKLY GROCERY-SHOPPING LIST

【推薦序1】 充滿愛的原味料理

陳美鳳
（知名主持人）

認識Kevin & Claire是在「美鳳有約」。這對夫妻真的好棒！好用心，好認真在作菜！我感覺他們把每次錄影都當作是最重要的一次，每次都能把食材的原味完整呈現卻又是色、香、味俱全，一吃就會上癮。

最重要的是，每道菜都充滿了愛心——「愛」是發自內心不斷的給予和付出，而不是斤斤計較如何「回收」，而這正是對這可愛夫妻的真實寫照！

謝謝Kevin & Claire讓「美鳳有約」的觀眾受益良多，也讓我品嚐到真正原味又美味的健康料理！

[Foreword]

Original Flavor Cuisine that is Full of Love

Chen, Mei-Fung
(Famous TV Host)

I met Kevin and Claire on the set of A Date with Mei Fung. They are a great couple! They put their love and heart into their cooking! I feel that they always treat each new filming as the most important one. And every time, they are able to present the original flavor of the ingredients in a way that is full of color, fragrance, and flavor. Trying their food will surely start an obsession.

The most important thing is that every dish is filled with compassion: a "love" that unceasingly gives from the heart without asking anything in return. This is the precise depiction of this loving husband and wife.

I thank Kevin and Claire for benefiting the viewers of A Date with Mei Fung in so many ways, and giving me a chance to experience the authentic taste of healthy original flavor cooking!

【推薦序2】 認真熱情的原味料理達人

陳凱莉
（前GOOD TV「健康新煮流」製作人、「家庭八點檔」主持人）

「堅持原味、鎖住營養」是KC夫婦掛在嘴邊最常說的口頭禪。

有一回，我們〈健康新煮流〉節目出外景到官田採菱角，Kevin就像尋獲寶藏一般，迫不及待現煮新鮮菱角給工作同仁品嚐。沒想到才一會兒工夫，只見他鍋蓋一蓋，算準時間就起鍋。雖然簡單又快速，但是，那一口香脆的菱角味道，對我而言，到現在依舊口齒留香。為什麼呢？因為Kevin精準計算蒸煮的時間，不但鎖住營養，也呈現菱角甘甜香脆的原味，有別於一般粉粉難嚥的口感，從此之後，我對菱角的印象真是大大改觀。

然而，大為改觀的事情還不只其一！在柿子園拍攝時，除了高掛樹上的脆柿成為食材焦點，連地上過熟的紅柿也是KC夫婦相中的目標。他們拾起過熟的新鮮軟柿，在小鍋裡以陽春作法攪拌熬煮，竟也作成了紅柿果醬。結果，我如法炮製回家帶孩子一起作，不但營造親子時刻，無形中也教孩子健康烹調的概念。

與KC夫婦互動，以「真實」兩個字來形容最為貼切。烹飪電視作業拍攝時，他們真是一點都不馬虎，幕前幕後認真的態度完全一致，並不因為入鏡與否就稍有怠忽。從清洗食材到最後盛盤的動作，總是口口聲聲說，必須以身作則並對觀眾有交待，光是這一點製作人就佩服得五體投地。

我想，原味烹調的專業態度，同時也是KC夫婦的生命態度，如此真實敞開的敬業精神，也正是您會愛上他們的原因。

備註： GOOD TV第15頻道
〈健康新煮流〉播出時間為：每週一至週四晚間8:30~9:00；每週五晚間8:00~9:00
PS：GOOD TV (Channel 15) - New Wave Cuisine show time :
Monday to Thursday 8:30-9:00 PM , Friday 8:00-9:00 PM

The Earnest, Sincere Experts of Original Flavor Cuisine

Kelly Chen
(Producer of GOOD TV New Wave Cuisine & Host of Family 8 PM)

"Insist on original flavor, locking in the nutrients," is a motto which Kevin and Claire take everywhere. Once, at an offset filming of New Wave Cuisine, Kevin could not wait to cook the fresh water caltrop for his colleagues as if he had found buried treasures. Without much ado, we saw him cover the lid of the pan and set the exact time to take it off the heat. Though it seemed quick and simple, the crunchy sweet flavor of the water caltrop stayed with me to today. Why? Because Kevin's perfect timing of cooking not only locks in the nutrients but also retained the original sweetness, which is different from the difficult to swallow and powdery taste that I usually experience with water caltrop. This meal completely changed my impression of water caltrop.

That was not the only change of perspective for me that day. When we were filming the persimmon farm, not only was the crispy fruit fresh persimmon on the trees a focal point of the show, even the overly ripe persimmon on the ground caught the couple's attention. They made persimmon jam by simply stirring and boiling down the ripe persimmon in a saucepan. Subsequently, I imitated the process at home with my kids, which not only gave me quality time with my family, but also provided a chance for me to teach them the concept of healthy cuisine.Interaction with the KC couple can be best described by the word, "sincerity." They take each filming seriously, keeping a genuine attitude whether on-screen or off-camera. They carefully explain each step, from cleaning the ingredients to presenting the food, to educate the viewers. Their diligence alone in practicing what they preach wins the admiration of the producer.

I believe their professional attitude in original flavor cuisine, concurrently also Kevin and Claire's attitude toward life, and such true expression of their professionalism toward work are precisely also the reasons why you will fall in love with Kevin and Claire.

【自序】

不能錯過的烹飪新浪潮～原味幸福料理

1997年，KC在最不景氣的時候成立，只為了推廣健康烹調，證明原味也可以很美味。

十年以來，歷經台灣921大地震、美國911事變、納莉颱風摧殘、SARS風暴、金融緊縮、投資外移⋯⋯KC從投資過度到負債經營，都不改心志，堅持到底就是要推動健康烹調！經濟的不景氣並未降低健康養生的影響力，反而興起更多的人追求健康風、樂活風；從有機飲食、NEWSTART生機文化、排毒餐到無油無水原味烹調蔚為風潮，一波又一波推動著忙碌的現代人找尋真正的健康出路。

去年Kevin和我有幸被加拿大政府邀請參觀當地的農漁牧業，也趁機觀摩不同的餐飲文化，其中有些餐廳就與當地的有機農場配合，使用當季的有機蔬菜作不同的有機料理，還會要求所有的廚師或廚助都要輪流到農場學習栽種和協助收成，主要是培養「尊重食材」的態度，才會愛惜食物並做出美味的料理；同時也透過特別的桌邊服務，讓用餐者慢慢享用這些精心料理的美食。

在加拿大農業及農產食品部專員夏文麗（左後一）的帶領下，進入蒙特婁美麗的楓樹林，認識楓糖漿的製作及享用主人御廚的楓糖變化餐。

Farm to Table 的慢食（Slow Food）哲學

「Farm to Table」的烹飪觀，開始風靡全球。除了強調使用在地百里內的食材，也強調用心烹調和用心享受的慢食文化，來取代現代速食觀。這樣的新飲食文化其實早已間接引進臺灣，有機食材和有機餐飲也跟著水漲船高，然而要在家中力行這樣的飲食觀，所有的食材都在百里內取得已不易，要全有機也很難；要全家一起力行於生活飲食中更難！

KC在電視上創作了數百道不同風味的健康料理，在餐廳接待過上千位體驗健康美味的客人，我們發現，簡單原味的烹調還是我們的最愛，也最讓客人回味！

New Wave Cuisine~ KC Natural & Heavenly Fare

Kevin & Claire

In 1997, a time of poor economy, KC was established in 1997, at a time of poor economy, to promote healthy cooking and to prove that the original flavor of natural food can also be a delicacy.

In the past ten years, we experienced Taiwan's 921 earthquake, 911 in America, typhoon Nari, SARS, economy crisis, investment offset... KC has been through over investment to running in debt, but has never changed its determination to promote healthy cooking. However, the slowdown in economy did not affect the increased awareness of health consciousness. Instead, more and more people follow this new trend of healthy lifestyle to achieve better health. Organic diets, NEWSTART vitality culture, expulsion of toxin meals and greaseless cooking are waves of the movement that encourage busy, modern people to find their way to good health.

Last year, Kevin and I had the privilege to be invited by the Canadian Trade Office to visit the local agricultural and fishing industry. We also took the opportunity to observe a different food culture. Some of the restaurants cooperate with organic farms and create many different recipes with seasonal organic vegetables. The chefs and kitchen assistants take turns to learn how to plant the vegetables at the farm and help with harvesting the crop. The main purpose is to develop an attitude of "Respecting the Food"- loving food in order to make delicious cuisine. Combined with special table service, diners can slowly enjoy the delicacies of these well-prepared gourmet foods.

Farm to Table - Slow Food Philosophy

The "Farm to Table" view of cooking is now a trend throughout the world. Not only does it emphasize using ingredients from local farmers (within 100 miles), it also emphasizes the culture of slow eating in contrast to the fast food mentality of the modern society.

只要照著KC料理精髓：選用天然新鮮未經加工的上等食材，透過正確的清洗、處理、冰箱儲存和健康的無油無水烹調（Greaseless and Waterless Cooking），要全家一起「吃出健康、吃出美味、吃出愛」真的一點都不難！

從原味的感動進入原味幸福料理

陸續收到讀者的反應：有人被書中的生命故事感動，有人被書中的冰箱管理吸引，有人渴慕擁有跟書裡一樣的開放式廚房⋯⋯來自各國各地的讀者或觀眾，都傳遞出一股透過改變烹調習慣，透過食物傳遞感情，改變了個人與家庭關係的震撼力。

- 中壢的吳先生帶著太太一起看節目、上課、看食譜做菜，半年後發現，太太的灰指甲改善，小孩排便順暢，也較不臭。
- 台中的素幸是標準的職業婦女，接觸KC前是全然的外食家族，連小孩早餐都是在隔壁的早餐店按月包伙的！但《原味的感動》吸引她作了改變，透過書開始看KC的節目並報名上了一堂無油無水入門課之後，就單純地按著書中的教導，一天一道菜的變換著，她的老公給她98分的鼓勵，小孩從營養不良開始恢復體重並愛上回家吃媽媽的愛心料理！
- 新加坡的Peggy按著書中所教的，每天早上一杯原味優格蔬果汁，原本退化的膝關節獲得改善；台灣的Tiffany，典型的摩登OL，也從原味優格蔬果汁改善了常應酬老公的便祕，還因此消了鮪魚肚。
- 美國的Tina透過網站收看好消息廚房，改變全家人的飲食觀；KC食譜更是她每餐必備的工具書。
- 一位大陸的忠實觀眾還曾說：「KC料理讓我們有機會接觸新的烹調觀，放下過去的傳統包袱，進入健康的新食代」。
- 「健康新煮流」的錄影團隊是道地的外食族，第一次錄影接觸KC料理就被深深吸引和懾服，徹底改變健康烹調觀，並愛上KC的原味料理。

「如何讓每個人都可以輕鬆下廚，愛上原味低脂烹調，落實健康

This new food culture has been in Taiwan for quite a while. The price of organic ingredients and food has also greatly increased. However, to live out this food philosophy at home, not only is buying ingredients difficult due to the long purchase distance from local farms, it is also unrealistic to use only organic products for all ingredients. In short, it is certainly quite an impossible mission to have the whole family adopting this lifestyle.

KC has introduced hundreds of different healthy recipes on our television program and served thousands of diners wanting to experience healthy, natural flavor-filled meals at our restaurant. We have discovered that simple and natural flavor cooking is not only our favorite, but it is also highly appreciated by our customers. As long as you follow the essence of KC cuisine – use natural, fresh, non-processed, quality ingredients with proper cleaning, preparation, and storage, and cook without grease and water – it is easy to have your whole family "eat for health, for flavor, and for love".

From Inspiration of Original Flavor to Natural and Heavenly Fare

We regularly receive responses from our readers: some are touched by our testimonies in the cookbook, some are attracted to the tips on refrigerator management, others yearn to have an open kitchen as described in the book. Readers and viewers from around the world express the impact of changing their cooking habits and learning to communicate love through their food.

· Mr. Wu and his wife from Chungli watch the show, attend cooking class, and practice the recipes in the cookbook . After six months, they see improvement to the wife's grey nails and the children's digestive systems.

· Mrs. Su from Taichung is a typical working mom who used to eat out every day and get up supply meals for her child at a breakfast store. After reading Inspiration of Original Flavor, she started watching KC's program and attended a class on greaseless and waterless cooking. From then on, she followed the cookbook and tried a new dish everyday with the support of her husband. Her child returned to a healthy weight and fell in love with his mom's cooking.

與愛的廚房。」是Kevin與我拍節目及出食譜的原動力，這樣的使命是建構在我們願意分享，每天每時每刻在廚房中經營生活的點點滴滴，幫助找不到答案和方法的人，進入原味烹調的異想世界。

當愈來愈多的新舊朋友反應KC出過的食譜各有不同的感動和烹調重點，有很多的資訊讓人想跟著落實，但如何集結重點，以簡易樸實的步驟，煮出健康味美的原味料理？如何讓那一群看了食譜、看了節目會很感動，卻無法實行的人產生馬上執行的行動力？

我們開始思索企劃一本能真正代表KC料理精髓，讓人可以馬上放棄傳統烹調習慣，邁向「健康新煮流」的關鍵食譜！

參訪全台最大的酪農區——柳營鄉，農會推廣股股長與可愛的田媽媽們興奮展示牛奶饅頭系列產品。

重新開始，從心出發

科技和先進社會所產生的精緻飲食文化，卻抹煞了原始的烹調美意：透過簡易又正確的調理方式，烹煮殺菌使食物更好消化和吸收以外，還要讓食物聞了有香氣、看了有食慾。但什麼才是健康的色香味呢？不再是講究刀工、擺盤及繁瑣的烹調，而是研究食物的屬性、大小、成熟度及時間掌控，在烹調時鎖住營養呈現出更美麗的原色、自然的菜香和原始的美味。

KC原味幸福料理就是在這樣的理念中成型，從調味料的認識、食材的挑選、基礎切功示範、控溫的技巧到一週菜單的計劃，只要按部就班，用心料理、用愛烹調，這本工具書可以幫助您真正的從內在的感動，進入實際的操作而產生滿足的幸福感！這也是KC跨越十週年，回應上帝滿滿的愛，送給所有追求健康與愛的廚房者最好的禮物！

013

轉化個人、家庭及社會健康觀的關鍵「食」刻

健康就是財富的觀念已深植人心，如何落實成為生活中不可少的習慣？從健康飲食切入是最實際也最容易的超連結。很多人無法理解，超越自己的積極觀，來自個人全方面的健康觀，也就是身心靈俱全的最佳寫照。一旦個人注重健康飲食，就會帶動全家健康吃的文

· Peggy from Singapore followed the instructions from the book and drank a cup of plain yogurt fruit and vegetable juice every morning and saw an improvement to her aging ankle. Tiffany from Taiwan, a typical modern office lady, also used the juice to fix her husband's problem with constipation and yet for this reason his tuna fish-like belly vanish.

· Tina from America watched the Good TV Kitchen program online through the internet and changed her entire family's view on food. She refers to KC's cookbook every meal she cooks.

· A faithful viewer from China once said, "KC's cooking gives us a chance to learn a new perspective on cooking and helps us to give up traditional habits and step into a new generation of healthy eating."

· The filming crew on New Wave Cuisine were used to eating out. After their first interaction with KC, they were deeply attracted and succumbed, radically changing their concept on healthy cooking and fell in love with original flavor cuisine.

How we are able to let everyone enjoy simple cooking, fall in love with low-fat original flavor cooking, and practicably build a kitchen of health and love is the motivation behind our TV shows and cookbooks. This mission is built upon our desire to share each moment of our experience in the kitchen with people who need guidance and answers to help them enter the fantastic world of original flavor cuisine.

New and old friends increasingly responded to our cookbooks by sharing their inspirations and takeaways. The wealth of information calls for a step of action; but how can we summarize the main points while keeping instructions to healthy cooking simple? How do we help the people who are inspired by the shows and the cookbooks but still hesitate to act? We therefore started planning for a fundamental cookbook that can truly capture the essence of KC cuisine, helping people to let go of outdated cooking habits and march toward the New Wave Cuisine of this crucial cookbook.

化，即便外食也會找健康餐廳而帶動餐飲健康風；當人民健康，社會成本就會降低，這正是政府在推動國民健康福祉的主要理念。只要健康飲食價值觀深入每個家庭時，人人都可以找到方法落實，從小小的個人開始作預防食療，加上媒體的協助及政府的帶動，全新的轉化力量會帶來美好又完全的影響力。

很多讀者與觀眾反應：真正的幸福是什麼？只要把菜煮得健康、原味又美味，家事及廚房管理得井然有序，已經很滿意，若再贏得心愛人的回眸一笑和一個大擁抱，就快樂的想飛起來呢！

這樣的要求很難嗎？

關鍵在於您願不願意轉化心境，放下過去的烹調習慣，踏上這波烹飪新浪潮，直接進入KC原味又幸福的健康新煮流！

推出《原味的感動》之際，我和Kevin也同心獻上感恩；如今《原味幸福料理》緊接在後，盼望透過這兩本書將同樣的祝福與恩膏降臨到每一位讀者，被原味的感動吸引時，願意改變傳統的烹調習慣，建立健康有愛的廚房，天天享受原味的幸福料理。

心動不如行動！唯有您打開KC食譜，力行其中的方法和步驟，健康烹調的種子就會在您家的廚房開花結果。

Technology and the modern society produced a delicate eating culture but also obliterated the beauty of natural cooking - a simple yet correct food preparation and a cooking process that besides killing harmful bacteria on food, make food more digestible, it also enhances the nutritional value while emerging the fragrance and make food more palatable and appetizing. A healthy combination of original color, fragrance, and natural flavor is no longer being particular about beautiful cuts, presentation and many nitty-gritty cooking methods, but a result of careful study of the ingredients' properties, size, desired doneness and cooking the ingredients with perfect timing to preserve maximum nutrition.

KC Natural and Heavenly Fare took form in this philosophy. Learning step-by-step, from understanding various seasonings, choosing the right ingredients, learning basic cutting techniques and controlling pan heat, to planning a weekly menu for the family, this fundamental cookbook act as an instrument to help you be inspired from within. So long as proceeding in order, get into practical action and truly cook with your heart and love, you will be filled with heavenly joy and happiness. This is also a way for us to celebrate as KC Kitchen stride over ten-year anniversary, responding God for His abundant love, and giving a gift to all of you in pursuit of true health and love.

A Crucial Time to Transform the Healthy Concept of Individuals, Families, and Society.

The concept that health is wealth is deeply rooted in people's heart. How can we make it into a daily habit and lifestyle? Starting from healthy eating is the easiest and the most practical way. Not everyone recognizes that positive thinking is a result of complete personal health, a balanced fullness in the body and soul. Once a person learns to appreciate healthy eating, the healthy culture will be brought to the entire family. Even when they eat out, they will look for healthy restaurants and thus promote the healthy trend in the industry. It is also the goal of the government to see the overall health of the people improve, which will then drive down the cost of health care. When every family is immersed with healthy eating values, the changes in individuals, with the help of media and government, can make a positive and absolute impact on society.

Many viewers and readers ask us: what is true joy? You will be well satisfied when you can cook healthy, naturally flavored food and keep an organized kitchen in your home. Then a simple smile and a hug from loved ones will make you feel like you are floating on air.

Is this too much to ask? The key is in your willingness to change, let go of outdated cooking habits, step onto this new wave cooking, and enter KC's natural and heavenly healthy new wave cuisine .

When Inspiration of Original Flavor was first published, Kevin and I together offered our thanks. Today, followed by the Natural and Heavenly Fare, we hope that through these two cookbooks, the same blessings, grace and annointing will fall upon each and every reader. When you are moved by the original flavor, be willing to let go of outdated cooking habits, build a healthy and lovely kitchen, and enjoy original flavor cuisine every day. Don't wait! Open KC's cookbook, step into action, and the seed of healthy cooking will bear fruit in your very own kitchen.

Origin Organic Farm in Delta, BC Canada.

本書使用說明

本書使用標準計量
Measurements used in this Book

量匙 | Measuring Spoons

● 一組有四種規格，有不鏽鋼、塑膠等材質，適用於食材量少時。
  Each set comes in four sizes, made of stainless steel or plastic. Measuring spoons are usually used for small quantity measurements.

1大匙=15 c.c.	1 tablespoon = 15 c.c.
1茶匙=5 c.c.	1 teaspoon = 5 c.c.
1/2茶匙=2.5 c.c.	1/2 teaspoon = 2.5 c.c.
1/4茶匙=1.25 c.c.	1/4 teaspoon = 1.25 c.c.

量杯 | Measuring Cups

● 一般有塑膠、不鏽鋼、玻璃及壓克力等材質，適用於食材量多時。
  For larger quantities, measuring cup measurements are used which are usually made of plastic, stainless steel, glass or acrylic material.

● 一般用的量杯標準容量是240c.c.
  The standard measuring cup usually used holds 240 c.c. or ml (milliliter)
  1 cup＝240ml
  4 cups (1 quart) ＝1 liter

● 本書用的是200c.c.不鏽鋼量杯。
  In this book, we use a 200 c.c. stainless steel measuring cup.

電子秤、量秤 | Electronic scale, traditional scale

- 用來秤量物品重量的工具。傳統式磅秤價格便宜，但是比較無法計算出微量材料的重量；而電子秤的精密計量至1公克的容量較精準，且體積小、操作容易、好收納，但價格較高。

  Scales are used to weigh ingredients. Traditional scales are less expensive, but it is harder to measure the exact weight of small amounts of ingredients. The electronic scale can measure as accurate as to the gram. It is smaller in size, easier to use and store, but higher in price.

1公斤=1000克	1 kilogram = 1000grams
	1 kilogram = 32 ounces (2 pounds) (1 ounce = 30 grams)
1斤=16兩=605克	1 catty = 16 liang= 605grams
1兩=37.8克	1 liang = 37.8 grams

特定材料份量指南 | Special Ingredient Quantity Guide

- 適量：表示依個人的口味增減用量，如黑胡椒、鹽等。

  A dash: amount added depending on personal taste, e.g. black pepper, salt, etc.
- 少許：表示只要略加即可，如蔥花、香菜末等食材。

  Just a small amount needs to be added, e.g. minced scallion or parsley.
- 檸檬皮1顆：表示是使用擠乾汁液的檸檬皮。

  1 lemon skin: lemon skin of lemon which juice has been squeezed out.

本書使用鍋具說明
Utensils used in this Book

　　以原味低脂烹調首重選對鍋具，才能鎖住營養到最高點並品嚐到原味與美味。健康又簡易的烹調關鍵，在於使用正確又好用的鍋具。天然的食材擁有不同的風味、維生素、礦物質、幫助消化的酵素及挑動食慾的色彩；然而很多的烹調方式往往剝奪了食物原有的天然好處。藉由獨特的蒸氣水膜設計鎖住食物本身的水分和營養烹煮食物，就不需使用太多的鹽、糖、油及其它的調味料；而挑對大小適中的鍋具料理食物（食材至少達2/3滿），可以保留較多的營養素及更多的美味。像我們夫妻一樣，熱愛並盡情享受健康烹調的人，會因為一個好的鍋具，極致超凡的烹飪表現而不由自主推薦選用；主要是好的鍋具可以料理出較健康又美味的食物——向食物借油，不添加多餘的熱量並鎖住營養到最高點，吃到更多的美味；向食物借水以保留更多的維生素及礦物質。

Selecting the Right Utensil for Maximum Nutrition and Flavor.

　　Good cookware is the key to healthier and easier cooking. Natural food contains abundant flavor, vitamins, mineral, digestive enzymes and color. However, many cooking methods can rob food of its natural qualities. Unique Water Seal locks moisture and nutrients inside to baste food in its own natural juices without adding excessive salt, sugar, oils or other sauces; selecting the appropriate size utensil (at least two-thirds full) will result in more nutritious and flavored-filled meals.

　　People like us who enjoy and love healthy cooking recommend the choice of a good cookware for its superior cooking performance and its healthy cooking features: "greaseless and waterless" cooking for nutritious, flavor-filled meals and maximum retention of the natural goodness without unnecessary calories.

標準平底煎炒鍋 | Standard Size Skillets

● 尺吋與用途　Size and Usage

約8吋、11吋及12吋，適用於無油
煎、烤魚肉類或少油煎蛋、豆腐及
炒菜俱佳。

The 8, 11 and 12 inch skillets are
suitable for grilling, pan broiling meats
and seafood or sauteing, pan-frying
vegetables, tofu and eggs.

鐵板電烤鍋 | Electric Griller

● 特色　Features

四合一多功能的電烤鍋，集烤箱、火鍋、飯鍋及鐵板燒等多功能的特色，可
放置在餐桌、吧台、客廳、院子，甚至可攜帶到親朋好友家一起做菜。它與
傳統烤箱不同處，在於其設計原理是靠包覆矽油的電熱圈，加上一體成型的
高碳鋼立體傳導的熱效應，可以達到均熱恆溫製作不同的料理。適用於需要
恆溫及控溫，如煎烤牛羊排肉類，海鮮料理如明蝦、龍蝦、螃蟹、花枝類，
燒烤糕餅、布丁、燉飯、火鍋及乾炒堅果花生等。

This is a 4-in-1 electric griller that functions as an oven, hotpot, rice cooker or
hot plate. It can be placed on the dining table or bar counter, in the living room
or yard or even taken to a friend's place to cook together. Its extraordinary
"oil core" construction with the heat control (unlike the top-of-the-range
fire), which can be precisely set at your desired temperature, provides gentle
and even heating at a constant temperature. It is especially suitable for recipes
that require high constant temperatures such as lamb chop, steaks (thick cuts);
seafood without fats like king prawns, whole lobsters, crabs, octopus; rice stews,

原味平底蔬菜或調味鍋 | Saucepan

● **尺吋與用途　Size and Usage**

有3/4公升、1公升、2公升。適用於快熟及少水烹煮的花果根莖類食物、清煮白米或五穀雜糧、煮少量的果醬、糖漿或魚肉醬汁等。

The 3/4, 1 and 2-quart saucepans are suitable for a pan full of quick-cook vegetables and roots with low-moisture cooking; plain and multi-grain rice; fruits such as strawberry sauce; or sauteing minced garlic and ginger for meats and fish sauce.

原味平底燉煮鍋 | Roaster/Stewing pot

● **尺吋與用途　Size and Usage**

有3公升、4公升、6公升、8公升及12公升，適用於煮湯、燒烤及同步乾煎、爆炒再進行燉煮、紅燒功能之料理、有一鍋兩用的雙重效果。

The 3, 4, 6, 8 and 12-quart stewing pots are suitable for soups, braises, stews and roasts. Its solar cap construction is perfect for pan broiling before stewing, so it requires no separate skillet.

如何檢測鍋溫
Testing for Pan Heat

若家中有好的不鏽鋼鍋具，可先將鍋具放在火爐上預熱，運用測水珠棒汲取少量的水，測量鍋中溫度。下頁是測量鍋子預熱由低溫到高溫的原則。

If you have some fine stainless steel cooking utensils in your kitchen, you can preheat them on the stove without grease and test the heat of the pans until water droplets "dance" when sprinkled in the pan. Below is the pan heat guide from low to high in consecutive order.

水珠慢跑 ▎ Water Droplets "Dance Slowly"

● 水珠顆粒較大，分散滾動率慢，消失時間較久。適合本身沒油的食材如蔬菜、太陽蛋、豆干及爆炒辛香料類。（水分較高如嫩豆腐、煎蛋等，水珠要比慢跑快一點。）

The water droplets are larger and "dance" around on the bottom of the pan in a slow manner. It takes longer for the droplets to disappear. This pan heat is suitable for cooking ingredients without fats such as vegetables, eggs (sunny side-up), bean curd, spices, pancakes, etc (Juicy foods like tofu, omelet, the water droplets should "dance" a little faster – termed "dance medium slow-quick".)

水珠快跑 ▎ Water Droplets "Dance Quickly"

● 水珠顆粒細小，分散滾動率快，消失時間較快。適合本身有油的魚肉類、雞排、牛排、羊排等。（有醃醬料之肉絲或肉塊如脫脂楓燒小排，水珠要比快跑慢一點。）

The water droplets are smaller and "dance" around on the bottom of the pan in a quick manner. The water droplets disappear quickly. This pan heat is suitable for cooking ingredients which come with fat, such as fish, meat, chicken filets, steaks or lamb chop etc. (Marinated meat such as defatted maple vinegar pork rib or shredded meats, the droplets should "dance" a little slower – termed "dance medium slow-quick".)

023

水珠跳動 ▎ Water Droplets "Bounce"

● 水珠顆粒細小，分散快速彈跳，消失時間較快。適合本身有油的內臟類如豬肚、大小腸等。

The water droplets are smaller and "bounce" around on the bottom of the pan in a quick manner. The water droplets disappear quickly. This pan heat is suitable for cooking animal parts such as pork stomach, intestines, hearts, brains, etc.

本書無油無水烹調法說明
Basics of Greaseless & Waterless Cooking

「**無油烹調**」是利用動物本身的脂肪來烹調，也就是向食物借油，達到烹調時只需用少數油或完全不用油。這個烹調方法的特色，在於它有效地運用了食物本身的油分，在導熱均勻的恆溫下乾爆或只使用少許的油，並逼出食物裡過多的脂肪。傳統的高溫油炸容易破壞食物的礦物質及維生素，導致食物縮水，特別是在高溫又開蓋的情況下，更容易使肉質縮小變硬，還可能要多油、多調味料烹調，提高熱量又不健康。

「**無水烹調**」是利用食物本身所含的水分來烹調，也就是向食物借水，達到烹調時只需用少許水甚至完全不用水。這個烹調方法的特色，在於它有效運用烹調時產生的蒸汽，以半真空的氣壓，迅速將食物半蒸半煮熟透。傳統的滾水烹煮，容易將礦物質溶解於水中，並流失食物的營養而破壞其原有的美味和顏色。

只要了解無油無水烹調的基礎奧祕後，就會像我們一樣，無可救藥地愛上健康烹調而不需要專業廚師的技能！再沒有一件事可以與我們在開放式的廚房所得到的刺激和樂趣相比！最重要的是，透過如此特別的烹調方式，不加鹽、不加糖、沒有油、沒有繁複調味料的情況下，仍然可以從新鮮天然的食材中，品嚐到最真實的原味和營養的幸福料理，而這正是過去品嚐的傳統料理所沒有的！

這是何等大的一個改變。它是不是真的如我們所強調的，還是更好呢？只要愛上無油無水烹調，就如同上了天堂般。這樣的期待，一定能為您帶來意想不到的「口福」。

"Greaseless cooking" means to cook with the fat that is naturally found in the meat. It is like borrowing the oil from the food rather than adding oil. It allows us to cook with little or no oil. It is an effective use of the oil within the food because we are able to "pan-broil" (without oil) or "pan-fry" (cook with a very small amount of oil) in an evenly heated pot to force out the grease from within the food. Deep-frying destroys minerals and vitamins, especially the high temperature combine with open cooking greatly adds to meat shrinkage. Adding excessive fats and oil, means added calories and it makes food seven times harder to digest.

"Waterless cooking" is to cook foods in their own natural moisture. It is like borrowing the water from the food to cook rather than adding water to the cooking process. Usually little, or no water needs to be added when the food is simmered in semi-vacuum heat. It also helps lock in flavor, color and vitamins when cooked in vapor steam to thoroughly baste foods. Boiling sterilizes food, however it dissolves water soluble minerals in the water and drain food of its natural goodness; it also destroys the flavor and color. Oxidation of food occurs when cooking without a cover, exposes food to the air, greatly reducing the quality of food. Traditionally, by stir-frying the vegetables, more oil will be absorbed, and excessive salt, sauces or MSG may be added.

If you understand the basics of greaseless and waterless cooking, you will be like us; and fall in love with healthy cooking and you do not need the expertise and skills of a professional chef. Nothing pleases us more than being able to recreate that excitement and fun in our open kitchen. Most important, you will discover that the true basic flavors and nutrients of fresh natural food without adding excessive salt, sugar, fats and sauces is much more than a heavenly fare than the truest traditional food you will ever taste.

What a wild gamble we take! Will it be as good as we emphasize, or even better? Anticipation tends to give your hopes such marvelous savor!

原味烹調小幫手：
蔥薑蒜酒汁的製作方法 Preparation of mixed juice of scallion, ginger, garlic and wine

蔥、薑、蒜、米酒依1：1：1：3等量，放入調理機打成泥狀，擠出成汁，裝入玻璃瓶後，冷藏約可保存3～4天。可用於去除腥味，適用於醃肉、醃海鮮或做料理。

Mixed juice of scallion, ginger, garlic and wine: Take equal portions of scallion, ginger, garlic and three times more rice wine and blend in a blender until mushy. Squeeze out the juice. It can be bottled in a glass bottle and kept in the fridge for 3~4 days. It is used to eliminate the unpleasant meaty scents and is suitable for marinating meat and seafood or as a seasoning in cooking.

PART 1
讓原味更美味的**基礎營養與採購學**
MENU PLAN and NUTRITION BASICS as
a PRIMER on TRUE NATURAL TASTES
to PERFECTION

小時候的我最喜歡拉著媽媽的手，一起到菜市場買菜，可以東看西看非常好玩，而且每樣食物看起來都很好吃，但是媽媽買菜大都是固定那幾樣食材。長大後自己上菜市場，買菜習慣也是跟媽媽一樣，直到認識Kevin一起研究健康烹調法，才警覺到，原來固定攝取的營養素有限，對增強身體的免疫力也會有限；因此在買菜之前作好事先的規劃，均衡分配採買多樣食材，可以均衡攝取到不同營養素，對身體健康也較有益處。就像我本身屬過敏體質，且容易發胖，改變飲食習慣後，新陳代謝功能及免疫力也都獲得改善。

真正成功的飲食計劃不僅安全又健康，而且可以享受食物原味的感覺；並且帶出均衡的飲食觀，可以有多重口味、不同顏色呈現，和豐富的烹調變化。愈來愈多人體會到健康與飲食息息相關，並了解到「藥補不如食補，預防勝於治療」。所以選擇一個真正可以持之以恆，持守到底的飲食計劃，一定要身體力行於每天的生活中，再加上學習如何吃的更恰當的心態來取代傳統的減肥觀時，一定可以輕鬆拋開過去不良的飲食習慣。

When I was a child, I loved going to the market with my mother. Hand in hand, we would look at all the interesting food available. Everything looked so appetizing, but my mother would always buy the same grocery items. When I grew up and did my own grocery shopping, I shopped in the same manner as my mother did until I met Kevin and researched healthy cooking methods together. This is when I realized that imbalanced diet may limit our immune system to function at full potential. Therefore, creating a menu plan to improve grocery shopping, and eating balanced, nutritious meals by selecting a variety of foods for consumption not only add nutritional value to our health, but also build up our body resistance to disease. Taking myself for example, I am allergic to certain food and easily gain weight due to improper diet. Once I changed my dietary habits, my metabolism and immune system both improved.

A truly successful eating plan is one that is safe, healthy and allows you to eat the foods you enjoy.It also offers a natural nutritional balanced diet that features a variety of flavors, colors and method of cooking.It is a paradox that more people are conscious of the relationship between diet and health, and food is a form of medication.It follows that what one chooses to eat ought to serve as a long term protection against illness.You must stay on the new plan until you have modified your harmful behaviors and gain control over eating.It is best to think of this activity as learning to eat properly rather than being on a diet.

Nutrition Basics starts with Understanding the Food Guide Pyramid

According to the new health trend of balanced diets suggested by doctors in recent years, the new Food Guide Pyramid is categorized into 6 main food group categories and expands them into the nutritional ingredients required in our daily diets today. The Food Guide Pyramid provides us with a basic guide to a balanced nutritional diet and the relevant breakdown of nutritional intake for each individual food group.

Six Main Food Group Categories of Nutritional Intake

1. Grains
· **Types of food** : rice, bread, cereal, multiple-grain, noodle, roots like potatoes, sweet potatoes and corn, etc
· **Recommended daily intake** : 6~11 servings per day

2. Vegetables
· **Types of food** : fresh leafy vegetables, rootstalks like celery, gourds, cruciferous vegetables, etc
· **Recommended daily intake** : a balanced diet of various kinds of vegetables, 3~5 servings per day

3. Fruits
· **Types of food** : various kinds of fresh fruit
· **Recommended daily intake** : 2~4 servings per day

4. Dairy Products
· **Types of food** : yogurt, milk, cheese, panna cotta, pudding, milk shakes, etc
· **Recommended daily intake** : 2~3 servings per day, preferably low fat dairy products

5. Meat, Fish, Eggs, Dry Beans and Nuts
· **Types of food** : fresh lean meat （beef, pork, chicken, duck）, fish, soy bean
· **Recommended daily intake** : 2~3 servings per day, preferably lean meat when selecting meat

6. Fats, Oils, Salt and Sugar
· **Types of food** : animal and vegetable fat, salt, sugar and nuts
· **Recommended daily intake** : use sparingly

基礎營養學，先認識金字塔飲食！

　　根據近年來醫學不斷提倡的均衡飲食新模式——即所謂「金字塔飲食」指標，依食物的營養特質分為六大類，可提供身體營養均衡方針及份量。

六大類飲食均衡攝取

第一類：五穀類

· **食物的種類**：米飯、五穀雜糧、麵食及根莖類的馬鈴薯、地瓜、玉米等。
· **建議每日攝取量**：建議每日3～6碗。

第二類：蔬菜類

· **食物的種類**：新鮮葉菜、根莖類如西洋芹、茄瓜類、十字花科類等蔬菜。
· **建議每日攝取量**：各類的蔬菜都要均衡的攝取，每人每天3碟。

第三類：水果類

· **食物的種類**：各種新鮮水果等。
· **建議每日攝取量**：每人每天2個。

第四類：乳製品類

· **食物的種類**：優酪乳、鮮奶、起司、鮮奶酪、布丁、奶昔等。
· **建議每日攝取量**：每人每天1～2杯，宜飲用低脂類奶品。

第五類：肉、魚、蛋、豆製品

· **食物的種類**：新鮮瘦肉（牛、豬、雞、鴨）、魚肉、黃豆、豆漿、豆乾、黑豆、蛋等。
· **建議每日攝取量**：每日4份，宜選擇食用瘦肉。

第六類：油、鹽、糖

· **食物的種類**：各種動物油及植物油、鹽、糖、堅果類。
· **建議每日攝取量**：控制攝取量。

Nutrition Basics

There are four major factors that influence our health: nutrition, exercise, the environment and heredity; we have the most control over our nutrition and the amount we exercise. Other than using the Food Pyramid as a guide for a daily balanced dietary intake, and on this account, choose a variety of food each week; also be flexible with your cooking methods. Plan a smart diet strategy for each week's menu in advance. When cooking, try to limit the amount of oil, salt, sugar and other seasonings, and focus on using natural spices. Consume less pickles, dessert and soft drinks; avoid deep-fried, highly seasoned dishes and over-processed foods with preservatives. Drink at least 6~8 glasses of water daily. Eat only 70% full; sustain a healthy and balanced nutritious diet to retard aging. Maintain an exercise routine for at least 30 minutes daily.

Basic Shopping starts with Creating a Menu Plan

Menu planning should take the interests of the whole into account, the age of family members and the amount of exercise we get. For example, if there are growing kids in the family, work meat and seafood as the principal food into different meals of the day; if planning for middle-aged couples, meat is not necessary on a daily basis since they do not exercise much, and incorporating a few vegetarian meals can be healthier for them. Planning a menu and shopping list makes good sense and is an efficient way for improving grocery-shopping to purchase the right quantity and proper foods. The weekly planned menu can be recycled bimonthly; the contents of which can be adapted seasonally or to the likes of the family members. The same ingredients can be cooked using different methods or seasonings or to match with different ingredients. For example, braised tofu can be modified to tofu with seafood, minced tofu topping, Japanese style tofu, etc; innovating diverse recipes and preparing your family for the blissful aromas and delights, which they will not be tired of.

Tips to Improve Grocery-Shopping :

Perfect Application of Time and Methods

Therefore, we would like to share with our readers a time-saving and

健康飲食的基本原則

影響健康的4大因素有營養、運動、環境及遺傳；而我們最能掌控的就是攝取的營養及運動量。此節可幫助我們除了每日依照「金字塔飲食」指標做為均衡飲食的原則外，每週最好選擇多種不同的食物做變化，且彈性調整烹調方式，聰明進行分配每週菜單，並限制油、鹽、糖及其他調味料的使用量，最好使用天然調味料，少吃醃漬品、甜食、汽水；避免食用油炸、佐料多又味道濃的菜餚、過度加工及有防腐劑等食品；每天喝6～8杯水，每餐堅持吃七分飽，均衡飲食抗衰老，每天持續做30分鐘的體能鍛鍊。

基礎採購學，先練習開菜單！

規劃菜單最好要依據家中成員的年齡做設計，例如家庭中有成長的小孩，可規劃每天一餐是肉食，下一餐則以海鮮為主食；而中年夫妻在菜單規劃上，並不一定要每天吃肉，因為運動量少，幾餐吃素反而對身體健康較有幫助。

針對家人需求規劃的菜單和採購單，有助改善採買的效率並買到品質好又合宜的食材；開出來的菜單，可以二個月輪流使用，每週隨季節或家中成員的喜好變化菜單內容。甚至相同的食材可利用不同的烹調法、搭配不同的食材或調味方式，例如紅燒豆腐，可改成海鮮豆腐、麻婆豆腐、日式和風豆腐等，讓家人吃不膩，每種食材皆可彈性變化不同的風味。

善用時間及方法，學習買好菜的撇步

因此我們想跟讀者分享一種最省時又簡單的方法──採買前先開立每週的菜單，採購時確實遵守「多樣選購，少量採買」的原則，即可達到均衡飲食的標準，又能每天像煮出美味的菜色。

我們的採購經驗是依據每週的菜單，選擇較新鮮種類多的攤販固定採買，並從買菜開始與老闆慢慢培養情誼，等建立好關係可進一步詢問老闆的聯絡電話，以後買東西在前一天訂貨，並說明採買的品種、分類法及切法（如小排、魚肉的厚薄及雞腿去骨）等需求，隔天取貨，除了節省時間，若是檢查不滿意還可以馬上更換，順便讓老

simple method—create a weekly family menu plan and shopping list before shopping and to really comply with the principle "shopping with variety and shopping in small quantities". On this account, we can at once attain a well-balanced nutritional intake, and yet resembles a professional chef who can cook delicious meals in an organized manner every day.

From our shopping experience, we shop in accordance with the weekly shopping list from regular vendors, that provide with more variety and freshness. Shopping from the regular vendors enables us to develop a relationship with the venders. Once familiar with the vendor, we can make further efforts to inquire the vendors' contact number and phone a day in advance to order the items we need. We can specify the assortment of food and how the food should be packaged, cut or other requirements (such as ribs chopped evenly, thicker cuts for sliced fish or meats, removing bone for chicken drumstick, etc), and then we pick up the items the next day. This saves time and in case something is not prepared as requested, we can ask for replacement right away. This also allows the vendors to be familiar with our great care for the foods' quality and requirements. Usually when vendors become familiar with their clients, they tend to reserve the best for their regular customers. Another bonus is when you need certain high-grade items, the vendors can order directly from their upper level supplier and saves you the time of bustling about various markets trying to locate that certain ingredient.

Therefore, start practicing to create a menu plan to help you be more organized in your grocery-shopping; there are many benefits: can avoid buying the same vegetables, maintaining a balanced diet for your family; control the amount of food you buy and save hard-earned money; when shopping at the market, you will not be scatterbrained and wonder what to buy or forget to buy something; it is a joyful event to plan menu with your family, you are bound to enjoy more blissful meals; it is an organized method of preserving your fridge in order, tidy and free of expired items. Henceforth, eliminate the hassle of cooking, cook simply and eat healthily.

闆了解你對食物的品質要求。通常與攤販老闆認識一段時日後，老闆大多會把最好的食物留給主顧客，而且還有一個好處，就是需要一些高級食材時，可以透過他們直接向上游廠商訂

開立每週菜單，才能準確又省時的買下最新鮮的食材。

購，而不用常常為了一個食材東奔西跑不同的市場。

所以練習開菜單，好處多多：可避買習慣的菜，維持全家人均衡飲食；可控制採買量；到市場不用左顧右盼或丟三落四的浪費時間；與家人共同開立菜單，每餐吃得更幸福；做事有條不紊，可避免冰箱囤積過期的食物。從此改變做菜的困擾，做得簡單，吃得健康。

從食材設計菜單，採買更有效率

那麼如何在採買及菜單上做出有效率的規劃呢？

以下列舉各攤位或店家販售的主要食材種類明細，供讀者參考，有助採買食物事半功倍。

豬肉
· 豬頸肉、梅花肉、中排、小排、豬絞肉、五花肉、豬腳、大骨、蹄膀肉、豬腱、小里肌肉、大里肌肉、胛心肉、豬肉火鍋片等。

牛肉
· 牛肩肉、牛肩胛、無骨牛小排、肋眼排、菲力排、牛里肌、肋排、沙朗、牛腩、前後牛腱、牛肉火鍋片、牛肚等。

羊肉
· 羊小排、羊肩排、羊腿肉、羊肉塊、羊肉火鍋片等。

雞肉
· 雞腿、雞翅、雞胸肉、雞里肌、雞爪、雞胗、粉肝等。

Efficient Shopping : Plan Menu According to Ingredients

How do we plan efficiently when it comes to menu planning and shopping? Below we have listed the assortment of the main items each vendor or store may sell as a reference for our readers. Hopefully, the list will be helpful, yielding twice the result with half the effort and save you time in your shopping process.

Pork Slab fat back, pork picnic shoulder, back ribs, spareribs, ground pork, slab bacon, pig's feet, big bone, hock, sinew, tenderloin, center loin roast, boston butt, sliced pork for steamboat, etc.

Beef Chuck flat iron, chuck flap, boneless short rib, rib eye steak, filet mignon, tenderloin, ribs, sirloin, rib finger, fore and hind beef shank, thigh, sliced beef for steamboat, tripe or stomach, etc.

Lamb/Mutton Lamb rack, shoulder lamb chop, thigh, lamb chunk, sliced lamb for steamboat, etc.

Chicken Chicken thigh or drumstick, chicken wings, chicken breasts, tenderloin, chicken feet, chicken's gizzard, chicken's liver.

Fish Salmon, yellow croaker, cod, tuna, milkfish, golden thread, saury, hairtail, grass carp, eel, larval fish, silver anchovy, yellow seabream, common carp, tile fish, promfret, sweet fish, Japanese seaperch, pike-eel, Japanese butterfish, groupers, tilapia, mullet, horse mackerel, marlin, snapper, cobia, yellow tail fish, scaled sardine, threadfin, etc.

Shellfish & Aquatic Products Lobster, slipper lobster, king prawn, grass shrimp, sand shrimp, stripe shrimp, spear shrimp, ocean white lu shrimp, raw prawn, blue swimming crab, serrated crab, horsehair crab, mangrove crab, Alaskan king crab, golden crab, crab legs, octopus, neritic squid, squid, calamary, cuttlefish, sea cucumber, stichopus, scallop clam, mussels, short-necked clam, raw oyster, ivory shell, small abalones, amazonian snail, hard clam, sea blubber, whelk, etc.

Vegetables **Leafy vegetables:** bok coy, bak choy, sweet potato leaves, Chinese broccoli, gynura, water convolvulus, Chinese spinach, Chinese kale, "huang kong" spinach, Chinese leek, chives, Chinese lettuce, spinach, crown daisy, lettuce, Chinese celery, lotus ferns, savoy, pucuk paku, baby cabbage, romaine lettuce, etc.
Cruciferous vegetables: cauliflower, leaf mustard, cabbage, broccoli, Chinese cabbage, etc. **Gourds:** pumpkin, loofah gourd, wax gourd, cucumber, big cucumber, bitter gourd, bottle gourd, eggplant, lady finger or okra, green pepper, bell pepper, corn, baby corn , water caltrop, zucchini, tomato, chayote, peng-hu luffa, green papaya, etc. **Rhizome/Roots:** white radish, turnip, rutabaga, horseradish, carrot, sweet potato, burdock, bamboo shoots, wild rice stem, asparagus, lily bulb, artichoke, taro, lotus

魚肉攤

· 鮭魚、黃魚、鱈魚、鮪魚、虱目魚、金線魚、秋刀魚、白帶魚、草魚、鱔魚、魩仔魚、丁香魚、赤鯮、鯉魚、馬頭魚、鯧魚、香魚、七星鱸、海鰻、肉鯽、石斑魚、吳郭魚、烏魚、竹筴魚、旗魚、鯛魚、海鱺魚、紅甘、青花魚、午仔等。

蝦、蟹、貝、軟枝水產

· 龍蝦、蝦菇、明蝦、草蝦、沙蝦、斑節蝦、劍蝦、海白蘆蝦、牡丹蝦、花蟹、紅蟳、青蟳、菜蟳、毛蟹、肉蟳、帝王蟹、黃金蟹、蟹腳、章魚、透抽、小卷、軟絲、花枝、魷魚、墨魚、海參、刺參、干貝、扇貝、蛤蜊、淡菜、海瓜子、生蠔、鳳螺、九孔、雪螺肉、文蛤、海蜇皮、新鮮螺肉等。

蔬菜

· **葉菜類**——青江菜、小白菜、地瓜葉、芥藍、紅鳳菜、空心菜、莧菜、油菜、皇宮菜、韭菜、韭黃、大陸妹、菠菜、茼蒿、萵苣、中國芹菜、水蓮菜、塌菜、水蕨菜、高麗菜苗、蘿蔓生菜等。
· **十字花科類**——花椰菜、芥菜、高麗菜、青花椰、大白菜等。
· **瓜果類**——南瓜、絲瓜、冬瓜、小黃瓜、大黃瓜、苦瓜、瓠瓜、茄子、秋葵、青椒、甜椒、甜玉米、玉米筍、菱角、櫛瓜、番茄、佛手瓜、澎湖絲瓜、青木瓜等。

澎湖絲瓜

· **根莖類**——蘿蔔、甜薯、蕪菁、大頭菜、辣根、胡蘿蔔、甘藷、牛蒡、竹筍、茭白筍、蘆筍、百合、朝鮮薊、芋頭、蓮藕、荸薺、洋蔥、馬鈴薯、山藥、蕗蕎、甜菜根、樹薯、西洋芹等。
· **芽菜類**——苜宿芽、綠豆芽、黃豆芽、蘿蔔嬰、碗豆苗、小麥草等。
· **豆菜類**——毛豆、四季豆、蠶豆、豇豆、豌豆、菜豆、長豇豆、荷蘭豆、甜豆筴等。
· **辛香配料類**——薑、大蒜、青蔥、青蒜、辣椒、九層塔、香菜、巴西里等。
· **野菜類**——山芹菜、過貓、山蘇、明日葉、山苦瓜、石蓮花、山茼蒿、龍鬚菜、川七、綠金針等。

巴西里

root, water-chestnut, onion, potato, Chinese yam, allium bakeri (leek bulb), beetroot, cassava, celery stalk, etc. **Sprouts:** alfalfa sprouts, mung bean sprouts, soybean sprouts, radish sprouts, pea shoots, wheat grass, etc. **Beans:** young soya beans, French beans, broad beans, cowpeas, peas, kidney beans, asparagus beans, pea pods, Chinese pea pods, etc. **Spices/Fresh Herbs:** ginger, garlic, scallion, leek, chilly, basil, Chinese parsley, English parsley, etc. **Wild vegetables:** honewort, fern, bird's nest fern, ashitaba, kakorot, houseleek, cotula hemisphaerica, gracilaria, madeira-vine, green day lily, etc.

Mushrooms Fresh black mushroom, shiitake mushroom, beauty white mushroom, king oyster mushroom, straw mushroom, abalone mushroom, tea tree mushroom, monkey head mushroom, Japanese buna shimeji mushroom, enoki or golden needle mushroom, button mushroom, golden cap mushroom, oyster mushroom, beech mushroom, Brazilian mushroom, matsutake mushroom, etc.

Soy Products Fresh bean curd, dried tofu or beancurd, tofu, shredded tofu, bean jelly, bean curd sheet, egg tofu, traditional tofu, etc.

Fruit White: honey peach, longan, mangosteen, Chinese pear, coconut, white peach, white grape, etc. **Yellow:** banana, orange, grapefruit, mango, pineapple, tangerine, loquat, passion fruit, pomelo, star fruit, durian, jackfruit, honeydew melon, kumquat, etc. **Green:** guava, kiwifruit, green apple, avocado, muskmelon, pear, sugar-apple, lemon, etc. **Purple:** purple grape, cherry, plum, blueberry, raspberry, sugar cane, mulberry, blackberry, etc. **Red:** apple, red grape, pitaya or dragon fruit, papaya, bell fruit or jambu, persimmon, tomato, watermelon, strawberry, rambutan, leechee, red quava, roselle, etc.

Traditional grocery store Egg, oil, rice, brown rice, glutinous rice, seasonings, multiple-grain, red bean, green bean, pearl barley, black & white sesame seed, dried bonito flakes, nuts, black soy bean, peanut, walnut, pine nuts, pumpkin seed, sunflower seed, dried silver anchovy, dried day lily, kelp, dried mushroom, black moss, seaweed bud, wheat, white fungus, black fungus, red cherry shrimp, carrageenan, dried cuttlefish, dried scallop, dried sea cucumber, shark's fin, dried oyster, mullet roe, dried sea shrimp, dried tiny shrimps, dried zbrias quagga, abalone, star anise, Szechwan peppercorns, cinnamon bark, dried red chilly, bamboo fungus, dried shitake mushroom, dried tofu skin, dried lotus seed, garlic, shallots, aged ginger, etc.

Supermarket Milk, soybean milk, egg, bread, cheese slice, noodle, rice flour noodle, seasonings, corn starch, flour, sweet potato flour, yogurt drink, yogurt, etc.

Bakery Whole wheat bread, French loaf, multi-grain bread, milk bread, squid ink bread, raisin bread, etc.

菇類

· 新鮮香菇、花菇、美白菇、杏鮑菇、草菇、鮑魚菇、茶樹菇、猴頭菇、柳松菇、金針菇、蘑菇、珊瑚菇、秀珍菇、真姬菇、巴西蘑菇、日本松茸等。

豆製

· 新鮮豆包、豆干、豆腐、干絲、涼粉、豆皮、雞蛋豆腐、傳統豆腐等。

水果

· **白色**──水蜜桃、龍眼、山竹、水梨、椰子、白甜桃、白葡萄等。
· **黃色**──香蕉、柳橙、葡萄柚、芒果、鳳梨、橘子、枇杷、百香果、柚子、楊桃、榴槤、波羅蜜、哈蜜瓜、金桔等。
· **綠色**──芭樂、奇異果、青蘋果、酪梨、香瓜、西洋梨、釋迦、檸檬等。
· **紫色**──紫葡萄、櫻桃、李子、藍莓、覆盆子、甘蔗、桑椹、黑莓等。
· **紅色**──蘋果、紅葡萄、火龍果、木瓜、蓮霧、甜柿、番茄、西瓜、草莓、紅毛丹、荔枝、紅芭樂、洛神花等。

雜貨店

薏仁

· 蛋、油、米、糙米、糯米、調味料、五穀雜糧、紅豆、綠豆、薏仁、黑白芝麻、柴魚片、堅果仁、黑豆、花生、核桃、松子、南瓜子、葵花子、丁香魚乾、金針、昆布、乾香菇、髮菜、海帶芽、小麥、銀耳、木耳、櫻花蝦、珊瑚草、乾魷魚、干貝、海參、魚翅、蠔乾、烏魚子、蝦米、蝦皮、扁魚乾、鮑魚、八角、花椒、桂皮、乾辣椒、竹笙、花菇、腐竹、乾蓮子、蒜頭、紅蔥頭、老薑等。

超市

· 牛奶、豆漿、蛋、麵包、起司片、麵條、米粉、調味料、太白粉、麵粉、地瓜粉、優酪乳、優格等。

麵包店

· 全麥土司、法國麵包、雜糧麵包、鮮奶土司、墨魚液麵包、葡萄乾土司等。

如何規劃每天三餐

　　有關每週開立菜單的食物搭配，我們會以少量菜色做烹調，務求均衡飲食。二個人的飲食每餐以一主菜、一配菜加一湯品或主食共三個菜色；三個人的飲食以一主菜、二配菜加一湯品或主食共四個菜色；四個人的飲食以二主菜、二配菜加一湯品或主食共五個菜色，菜量依人數增加的同時，菜色可依家人喜好增加共五至六道菜。

KC夫婦　一周採買規劃示範表

	星期一	星期二	星期三
早餐	原味優格蔬果汁 （芭樂、香蕉、萵苣、 小黃瓜、優格、水） 五穀雜糧饅頭 或南瓜全麥饅頭	原味優格蔬果汁 （奇異果、西洋梨、西洋芹、 蘿蔓、優格、水） 水煮黃地瓜	原味優格蔬果汁 （木瓜、蘋果、胡蘿蔔、 紅萵苣、優格、水） 水煮糯玉米
午餐	蘿蔔封肉或香茅燉雞 （P.190或P.112） 脆炒皇宮苗（P.170） 蕎麥糙米飯	西芹番茄炒墨魚（P.146） 或橙汁軟絲（P.148） 冰鎮柴魚菠菜（P.168） 紅麴米飯	紅酒燉牛腩（P.190） 烤全麥雜糧麵包
晚餐	五穀營養粥 洋蔥煎蛋（P.138） 干炒四季豆（P.184）	滑蛋絲瓜（P.158） 紅蔥酥乾拌麵線 柴燒杏鮑菇（P.180）	鮮味蝦或彩椒蘆筍燴鮮貝（P.152或P.194） 魚香茄子（P.162） 霧峰香米飯

　　每天三餐以**早餐吃好（代表金）、午餐吃飽（代表銀）、晚餐吃少（代表銅），每餐堅守七分飽為原則**；每週則可依小週末（星期三）、大週末（星期六）及假日設計較豐盛或特別餐，給家人不同的感受；星期五則適合以消化冰箱剩菜以利隔日重新採買，如少許甜椒、少許洋蔥、少許肉絲等，也可燴炒成「鐵板蔬菜豆腐」或「和風銀芽肉絲」之類的菜色。

星期四	星期五	星期六	星期日
原味優格蔬果汁（紫葡萄、青蘋果、紫高麗菜、苜蓿芽、優格、水）水煮馬鈴薯	原味優格蔬果汁（紅火龍果、香瓜、去皮馬鈴薯、大黃瓜、優格、水）烤全麥雜糧麵包	原味優格蔬果汁（鳳梨、柳橙、牛蒡、山藥、優格、水）水煮蛋	原味優格蔬果汁（草莓或藍莓、酪梨、荸薺+碗豆嬰、優格、水）水煮紅地瓜
脫脂楓糖醋燒小排（P.120）五穀雜糧飯秋葵塔（P.186）	和風銀芽肉絲（P.122）胡瓜蝦仁炒蛋（P.160）糙米稀飯	原味什錦炒米粉（P.204）楓燒雞翅（P.114）全麥無酵鬆餅（P.208）	黃薑脆皮雞腿（P.110）南瓜薯泥沙拉（P.154）彩虹菜鍋（P.174）
椒麻黃魚（P.132）乾煎鮭魚香椿醬炒飯（P.202）蔬菜瘦身湯	鐵板蔬菜豆腐（P.140）蒜香木須高麗菜（P.176）糙米飯	椒鹽烤無骨牛小排（P.128）或椒鹽烤明蝦（P.142）芥藍炒蘑菇（P.172）蘿蔓生菜紅豆糯米甜湯（P.188）	奶香淡菜盅或原味鮮貝盅（P.144或P.196）法國吐司麵包或原味牛肩肉骨茶湯（P.126）義大利細麵或越南河粉

Planning Daily Meals

When planning our weekly menu, we aim to prepare fewer dishes, yet seek to achieve a well-balanced nutritious diet.   When preparing the daily main meal (either lunch or dinner) for a diet for two, the meal includes one main dish, one side, and a soup or a staple food totaling three dishes; diet for three, include one main dish, two sides, and a soup or a staple food totaling four dishes; diet for four, include two main dishes, two sides and a soup or a staple food totaling five dishes. As the number of people increase, the food quantity should increase so that there are enough portions for each one, suggesting a total of five or six dishes according to the likes of the family members.

Kevin & Claire's Weekly Menu Plan (Sample)

	Monday	Tuesday	Wednesday
Breakfast	Natural Yogurt Fruit and Vegetable Juice (guava, banana, butterhead lettuce, cucumber, yogurt, water) High-Fiber Nutritious Bun or Pumpkin Whole Wheat Bun	Natural Yogurt Fruit and Vegetable Juice (kiwi, European pear, romaine, celery stalk, yogurt, water) Boiled Yellow Sweet Potato	Natural Yogurt Fruit and Vegetable Juice (papaya, apple, carrot, red Batavia, yogurt, water) Boiled Corn
Lunch	Braised Pork with Carrot and White Radish or Chicken Stew with Lemon Grass, Crisp "Huang Kong" Spinach, Buckwheat Brown Rice	Pan-Broiled Squid with Celery and Tomatoes Stir-Fry or Lemony Orange Glazed Squid, Iced Spinach Salad with Bonito Flakes, Red Yeast Rice	Red Wine Beef Stew, Tofu & Egg Soup, Whole Wheat Toast
Dinner	Multiple-Grain Porridge, Onion Omelet, French Beans with Bean Curd Stir-Fry	Loofah Gourd in Egg Smoothie, Dry Mixed Thin Noodles with Fried Shallots, Glazed King Mushrooms with Bonito Flakes	Shrimp Delight or Pan-Seared Scallops with Colorful, Peppers and Asparagus Saute, Teppanyaki Eggplant Topping with Saucy Minced Pork, Taiwan Wufeng Fragrant Rice

Of the three meals within the day, breakfast should be the best in quality (represent gold), lunch the most in quantity (represent silver), and dinner the least in quantity (represent bronze). For every meal, avoid eating until you feel stuffed, consume only 70% full. Every week, you can normally gather different recipes for your family and flood your house with blissful aromas by preparing special and sumptuous meals every Wednesday, Saturday and even on holidays. On Fridays, it is a good opportunity to finish the left over in the fridge so it is more convenient to buy new ingredients the next day; For example, you can make dishes like "Vegetarian Tofu Teppanyaki" or "Shredded Pork and Sprouts" with the little amount of the left-over of the bell peppers, onions and shredded pork.

Thursday	Friday	Saturday	Sunday
Natural Yogurt Fruit and Vegetable Juice (purple grape, green apple, purple cabbage, alfalfa sprout, yogurt, water) Boiled Potato	Natural Yogurt Fruit and Vegetable Juice (red pitaya, muskmelon, peeled small potato, big cucumber, yogurt, water) Whole Wheat Multi-Grain Bread Toast	Natural Yogurt Fruit and Vegetable Juice (pineapple, orange, burdock, Chinese yam, yogurt, water) Boiled Egg	Natural Yogurt Fruit and Vegetable Juice (strawberry or blueberry, avocado, water chestnut, pea sprout, yogurt, water) Boiled Red Sweet Potato
Defatted Maple Vinegar Sparerib, Multi-Grain Rice, Towered Gumbos	Shredded Pork and Sprouts, Big Cucumber with Shrimp Omelet, Brown Rice Porridge	Natural-Flavored Assorted Vegetables Rice-Flour Noodles, Maple-Barbecued Chicken Wings, Healthy Whole Wheat Pancakes	Crispy Turmeric Chicken Drumstick, Mixed Pumpkin Potato Salad, Rainbow Vegetarian Skillet
Pan-Broiled Salmon or Szechwan Peppercorn Yellow Fish, Stir-Fried Fragrant Cedar Garlic Rice, Vegetarian Dietetic Soup	Vegetarian Tofu Teppanyaki, Cabbage with Black Fungus & Garlic Stir-fry, Brown Rice	Grilled Peppercorn Boneless Short Ribs or Grilled Peppery Salt King Prawns, Cabbage Mustard with Button-Mushroom Stir-Fry, Glutinous Red Bean Dessert	Grilled Creamy Mussels or Simple Clam Delight vs. French Toast or Chuck Flat Iron Beef Broth vs. Spaghetti or Vietnamese Noodles

041

PART 2
讓原味更美味的
食材處理、刀工與烹調要領
Correct Food Preparation,
Cutting Instructions and
Healthy Cooking Tips
to Enhance the Natural-Flavor.

原味烹調的特色在於烹煮時盡量以原形或大塊烹煮，小塊分食；其最高的技巧是推翻傳統烹調觀念，在料理的過程中，從食材的處理、刀工到料理上桌，更簡單、更健康，更重要的是，吃起來比傳統的煮法更美味。

對於偏愛重口味的族群，尤其是認定健康的食物不好吃，且無法達到期望值的人來說，本書料理鐵定可以滿足您挑剔的味蕾。

只要逐漸地習慣無油無水的健康料理，一定會對每樣食材的原汁原味上癮，並且被混合著特殊鮮美口感及充滿活力的色彩所著迷！

The characteristic feature of natural-flavor cooking is to cook the natural ingredients in whole and is best to cut after cooking; the high technical skill is that though it revolutionize the traditional cooking concept, however the whole cooking process from food preparation to cooking is much easier and healthier, and most important it taste better than traditional food.

For spicy food lovers, especially those with the idea that healthy food is not delicious and unable to meet their expectations, the dishes in this recipe book is sure to please your finicky palates.

As you gradually become accustomed to these "greaseless and waterless" healthy dishes, your taste buds will definitely be addicted to the original flavor-filled taste of each ingredient, in addition to the blending of their succulent flavors and delicate textures of natural food.

保持食物營養的原味烹調基礎

　　保留食物原味的關鍵第一步，必須先了解食材前處理。食材處理方式正確，可減少農藥殘留，而且魚不腥、肉沒有異味，青菜更鮮脆，還能維持食物的新鮮度，攝取更多的營養素。

　　原味的烹調講究健康又美味，因此在這個單元中，我們將介紹一些烹調方便的基本器具、鍋具、砧板及刀具的清洗和保養，幫助您快速掌握食物保留原味的關鍵細節，保證在短短的時間內就愛上如此健康、簡單又美味的烹調方式。

　　The first key step to preserve the natural flavor of food is the awareness of correct food preparation. Using the correct method to prepare the food can reduce the remaining pesticides, as well as eliminating the unpleasant fishy and meaty scents, and also vegetables can be crispier. Further more, the freshness of the food will remain, so one can absorb more nutrition.

　　The natural-flavored cooking strives for health and taste; therefore in this section, we will introduce some basic utensils, cutting boards, knives and their maintenance and upkeep. These details will quickly enable you to understand the crux of natural-flavored cooking. Within a short time, you will fall in love with this healthy, easy and delicious cooking – I guarantee it!

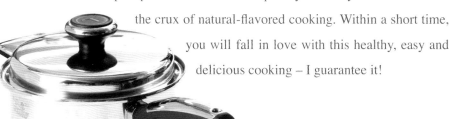

鍋具的選擇與清洗，是健康烹調的第一步。

保持食物原味的食材處理法

Correct Food Preparation for Natural-Flavored Cooking

魚 ▍ Fish

1. 徹底刮除魚鱗 。
2. 清除魚腹內臟時，魚腹不要開太大。
3. 利用粗鹽、檸檬皮洗魚去腥，再用冰鎮的過濾水沖淨後濾乾。

1. Remove the scales thoroughly.
2. Give a small cut along belly to gills and remove the innards and gills.
3. Eliminate the fishy scent by rubbing it with coarse salt and lemon skin. Wash clean with lemon skin in cold icy filtered water, drain.

· **貼心叮嚀**：魚腹若開太大，煎魚翻面不美；或是在魚腹兩面劃刀，煎煮時則容易流失原汁。以檸檬皮和粗鹽清洗食材，有助防腐和殺菌。

· **Remarks**：The fish may not be turn over beautifully if the cut along belly is too long; do not cross cuts on both sides of the fish to preserve the natural juices during pan grilling. Lemon skin and coarse salt is good for preservation and sterilization purposes.

豬肉 | Pork

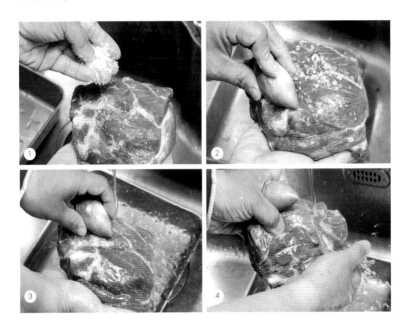

1. 撒適量的粗鹽。

2. 用檸檬皮塗抹表面去腥味。

3. 同步用過濾水及檸檬皮搓洗肉塊。

4. 用冰鎮的過濾水沖洗乾淨並濾乾，再依料理需求切割大小，可保留較多的原汁原味。

1. Sprinkle an appropriate amount of coarse salt.

2. Rub the lemon skin on the pork to eliminate the meaty scent.

3. Rub with lemon skin while rinsing with filtered water.

4. Wash it clean with cold icy filtered water, drain; the pork should not be cut into smaller pieces prior to cleaning, to preserve the natural sweetness and juices.

· **貼心叮嚀**：原味烹調肉類的大小塊，必須配合同道料理的其他食材屬性，所以最好是買回家自行切割，才能同步烹煮至剛好熟透。

· **Remarks：** For natural flavor cooking, the size of the meat is cut in proportion to the other ingredients of the same dish according to their nature, so it is best to cut personally to allow every ingredient to be just cooked.

雞肉 | Chicken

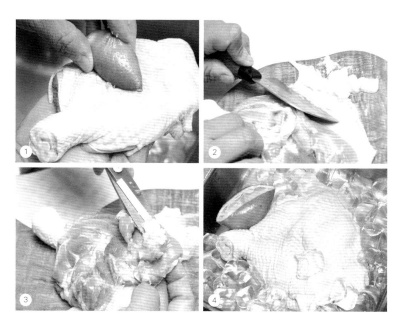

1. 用檸檬皮及粗鹽逆向搓洗毛
   細孔，去腥味及表皮污垢。

2. 刮除皮下脂肪。

3. 修剪瘦肉內的脂肪。

4. 同步用檸檬皮及冰鎮的過濾
   水沖洗乾淨並濾乾。

1. Wash the chicken by rubbing lemon skin
   and coarse salt against the pores, to
   eliminate the meaty scent and to remove
   any dirt from the skin.

2. Scrape out the fat under the skin.

3. Trim off the fat from within the lean
   meat.

4. Wash clean with lemon skin in cold icy
   filtered water, then drain.

- **貼心叮嚀**：為了保留雞肉的原汁原味，料理雞肉前，要修剪多餘的脂肪，就不需
  靠過多的調味料去除肉腥味及油膩感。

- **Remarks**：To savor the natural essence of the chicken, it is necessary to trim off
  excess fat before cooking so that the use of excessive sauces to eliminate the
  greasiness and unpleasant fatty texture can be omitted.

蝦 | Prawn

  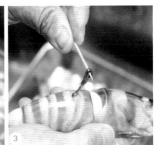

1. 剪除蝦腳後，較容易清洗；
   煎烤時，方便貼鍋站立，
   避免蝦身捲曲。

2. 剪除蝦頭尖角。

3. 將蝦平放在手上，再用牙籤
   自蝦背的第二節挑出腸泥，
   同步用檸檬皮及粗鹽以冰鎮
   的過濾水洗淨並濾乾。

1. Trim off the legs for easy cleaning of dirt and also to allow prawn to stand firm, preventing it from curling during pan grilling.

2. Trim off the sharp tip of the prawn head from getting hurt while eating.

3. Use a toothpick to devein prawn at the second section on the back, by placing the prawn slightly flat on the palm so that you only need to pierce once to remove the intestinal vein at a stretch. Wash clean with lemon skin and coarse salt in cold icy filtered water, drain.

· **貼心叮嚀**：原味烹調高級的深海明蝦時，最好是帶殼料理，且不要在蝦腹或蝦背切割，避免流失其原有的營養素及甜分。

· **Remarks**： Especially for high grade ocean king prawns, it is best to retain the shell and not to make any cut half way into the belly to prevent curling or cut at the back to devein, in order to preserve the natural sweetness and nutrients.

頭足類 | Cephalopod

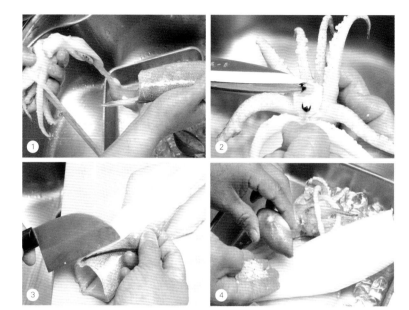

1. 頭與身體先剝開並沖洗身體內部，同時去除頭部雜物。

2. 剪除眼部硬殼。

3. 去除表面皮膜。

4. 同步用檸檬皮及粗鹽以冰鎮的過濾水洗淨並濾乾。

1. Separate the head and body. Wash the insides. Discard the greyish bag and all material near the head.

2. Pull out the cornea.

3. Peel off the membrane.

4. Rub with lemon skin and coarse salt. Wash clean with lemon skin in cold icy filtered water, drain.

· 貼心叮嚀：交叉斜切刻花再切成小塊烹煮時，需要高溫及多油爆炒，容易流失原汁原味；將身體與頭完整地放入平底鍋中無油煎烤，既健康又容易烤成金黃色，不需加硼砂，口感吃起來一樣鮮嫩香Q。

· Remarks： Do not slit the body or make diagonal cuts on inside surface to form diamond-shape cuts and then cut into small pieces, as more oil and higher heat is required to fry the squid. By pan-broiling the whole squid without oil, it is not only much more healthful but allows it to be browned easily; it also enhances the succulent flavor and chewy texture of the squid without having to add sodium borate.

豆腐 | Tofu

1. 脫模放入容器。

2. 以過濾水清洗，不要直接沖洗而弄破豆腐。

3. 濾乾水分。

1. Empty out and place into a container.

2. Wash clean with filtered water; do not add water directly into tofu to avoid being smashed.

3. Drain dry.

- **貼心叮嚀**：如果買傳統的板豆腐，當天未馬上料理，最好放在容器中以過濾水淹過豆腐；原味烹調豆腐，不要切太薄，至少厚約3公分，以少油加蓋的方式清煎。可保留豆腐表面較平整，並鎖住豆腐原汁，吃起來更香濃，有布丁或豆花的口感。

- **Remarks**：For traditional Chinese tofu, if not used the same day of purchase, immersed in filtered water to retain the fullness of the tofu. Do not slice the tofu too thin, at least 3 cm. thick, to preserve the smooth and delicate pudding-like texture. Frying in thick slices using minimum grease under closed cooking is also able enhance its fragrance.

菇類 | Mushrooms

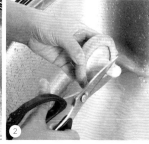
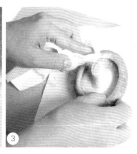

1. 用過濾水沖洗香菇的蕈褶。

2. 修剪少許香菇蒂，並注意不要低於香菇的帽緣。

3. 取紙巾輕壓香菇擦乾水分。

1. Rinse and wash away dirt and particles with filtered water.

2. Trim off a section of the stem end, leaving the remaining part of the stem to be extended slightly over the whole mushroom cap.

3. Drain, pat dry with a paper towel.

· **貼心叮嚀**：烹煮新鮮香菇以整朵的方式料理，可保留較多的營養素；稍長的蒂頭則可保護香菇的帽緣。

· **Remarks**：To retain maximum nutrition, the mushroom is wholly preserved; the extended part of the stem is to protect the edge of the cap.

貝類 | Shellfish

1. 用棕櫚刷刷洗表面污垢。

2. 放入容器，加粗鹽及鐵湯匙。

3. 倒入過濾水浸泡數小時，放在陰涼角落吐沙，用過濾水沖淨並濾乾。

1. Scrub clean with a palm brush.

2. Place them in a bowl with coarse salt and metal teaspoons.

3. Immersed with slightly chilled filtered water and let soak for a few hours in a cool dark corner, to release the sand. Change the filtered water once and drain.

葉菜類 | Leafy vegetables

1. 以大量清水沖洗葉莖及根部殘留的泥土，再切除少許的根部。

2. 依生長方向直立浸泡過濾水20分鐘，換水再泡20分鐘。

3. 濾乾水分並放入有蓋的保鮮盒，移入冰箱冷藏保鮮。

1. Rinse the stem and root end with a large amount of water, to flush away the remaining soil. Cut a little part of the root end.

2. Soak vegetables vertically（mimics the way vegetables grow）in the filtered water for 20 minutes. Change the water once and then continue the soaking process for another 20 minutes.

3. Drain, store in sealed container and cool in the fridge.

· **貼心叮嚀**：使用過濾水浸泡各類蔬果比用自來水浸泡保存的久；因為自來水中的氯及其它化學成分，會破壞蔬菜中的維生素及礦物質。在蔬菜根部切新傷口有助其快速吸收過濾水並且推擠出農藥、其他的化學成分及灰塵於水中。葉菜類要避免與其他較重的蔬果如白蘿蔔、紅蘿蔔一起浸泡才不易被壓傷；以大的不鏽鋼鍋儲存蔬菜較容易吸冰，並有效保存其新鮮度和脆感。

· **Remarks**：It is important to use filtered water to soak vegetables and all kinds of food. Vegetables will decay sooner if they have been soaking in tap water, as the chlorine and other chemicals in tap water will destroy the vitamins and minerals in the vegetables. Making a new cut in the root helps the vegetables to absorb filtered water faster to the fullest and thus releasing the pesticides, other chemicals and dust into the water. To avoid bruising the leafy vegetables, do not soak with other heavy vegetables such as radish and carrot. Store the vegetables preferably in a big stainless steel pot as it can absorb the coolness more rapidly and better retain it; thus preserving the freshness and crispness of the vegetables more effectively.

根莖類 Rhizome

1. 用棕櫚刷輕輕刷洗表面汙垢。

2. 切除莖及少許的頭部，帶皮浸泡過濾水20分鐘，換水再泡20分鐘。

3. 濾乾水分並放入有蓋的保鮮盒，移入冰箱冷藏保鮮。

1. Scrub lightly with a palm brush.

2. Remove the stem and cut a slice off the head of radish or carrot, do not peel, and soak in filtered water for 20 minutes. Change the water once and repeat process.

3. Drain, store in sealed container and cool in the fridge.

· **貼心叮嚀：**去除蘿蔔根莖類的莖，可避免其繼續吸取根部的養分而逐漸空心老化。其它的地下根莖類如地瓜、芋頭、馬鈴薯、洋蔥、老薑、蒜頭、紅蔥頭等，可直接放在陰涼通風處約一星期，之後依同樣方式清洗，並放入冰箱儲存以避免乾裂老化。

· **Remarks：**Discard the stem of radish or carrot to prevent the stem from absorbing its moisture from the root, thus causing hollowness of the cavity. For certain rhizome such as sweet potato, yam, potato, onion, aged ginger, garlic, shallot, put them in a cool, ventilated place, preferably for 1 week, and subsequently to apply the above stages of washing prior to storing in the fridge. It should not be stored too long outside the fridge, or they will shrink and get withered.

1. 用軟毛刷清洗表面汙垢。
2. 浸泡過濾水20分鐘後，換水再泡20分鐘。
3. 濾乾水分並放入有蓋的保鮮盒，移入冰箱冷藏保鮮。

1. Use a soft brush and scrub clean.
2. Soak in the filtered water for 20 minutes, change water once and repeat process.
3. Drain, store in sealed container and cool in the fridge.

· **貼心叮嚀**：帶皮直接食用的水果如新鮮莓果、葡萄、小番茄等，應分開浸泡及存放才，不會被壓傷。
· **Remarks**：To avoid bruising, the soaking process for fruit such as fresh berries, grapes, small tomatoes, etc, should be separated.

原味烹調最方便的基本器具

Most Convenient and Basic Tools for Natural Flavor Cooking

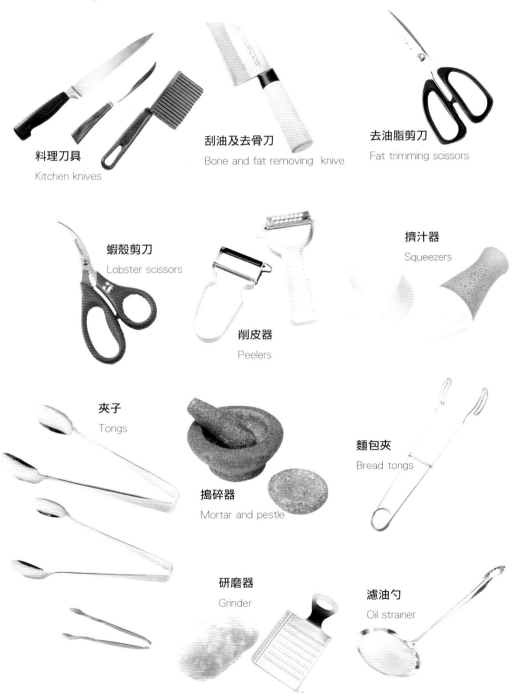

料理刀具
Kitchen knives

刮油及去骨刀
Bone and fat removing knive

去油脂剪刀
Fat trimming scissors

蝦殼剪刀
Lobster scissors

削皮器
Peelers

擠汁器
Squeezers

夾子
Tongs

搗碎器
Mortar and pestle

麵包夾
Bread tongs

研磨器
Grinder

濾油勺
Oil strainer

055

木匙
Wooden spatula

醃肉保鮮盒
Marinating meats glass container

隔熱墊
Heat insulation stand

切肉砧板
Meat cutting board

抹布（可以顏色區隔鍋具與檯面專用）
Rag （separate color for different uses）

計時器
Timer

熟食砧板
Cooked food cutting board

水果砧板
Fruit cutting board

平底鍋煎鏟
Stainless steel slice for flat bottom pan

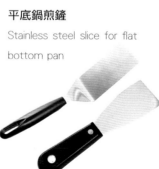

測水珠棒及食物刺針
Tester for pan heat and cooking pin

不鏽鋼製備料容器
Stainless steel food preparation tray

醃海鮮盒
Marinating seafood glass container

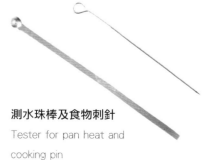

烹飪器具基礎的清潔法
Basic Cleaning Method for Cooking Tools

切菜板 | Cutting Boards

 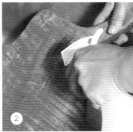 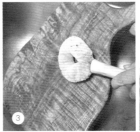 

1. 用刀刮除表面汙垢。
2. 噴鹼水去油及異味。

3. 用棕櫚刷刷洗乾淨。
4. 用熱水洗淨，瀝乾，置於通風處。

1. Scrape out the remains with a knife.
2. Spray washing soda water to remove stains and odor.
3. Scrub with a palm brush.
4. Rinse clean with hot water, drain and keep dry in ventilated place.

刀具 | Knives

  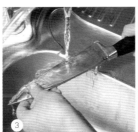 

1. 噴鹼水去油及異味。

2. 用海綿刷洗，偶爾用不鏽鋼清潔膏保養亮面。

3. 用熱水洗淨。
4. 瀝乾水分，放入刀架以免刀尖受損。

1. Spray washing soda water to remove stains and odor.
2. Use a sponge and scrub clean, occasionally polish with stainless steel cleaner for a shiny surface.
3. Wash clean with hot water.
4. Drain, keep dry and store in knife holder, to protect the knife from tipping up or broken at the tip.

經常保養鍋具守則
Practice Regular Cleaning for Pots

1. 用微濕的海綿沾不鏽鋼清潔膏及洗碗液，以劃圓圈方式刷洗鍋蓋及鍋具的亮面。

2. 以菜瓜布沾不鏽鋼清潔膏及洗碗精，刷洗鍋蓋內面及鍋底。

3. 最後用菜瓜布，刷洗鍋具最髒或油膩的內部。

4. 用熱水將裡外沖淨並瀝乾。

1. Smear the stainless steel cleaner and dishwashing liquid on the soft side of the slightly wet sponge; clean in circles the top of the cover and the exterior of the pan (the more shiny surfaces).

2. Smear the cleaners on the rough side of sponge to clean the base of the pan and the inside of the cover.

3. Lastly, scrub the interior that is the dirtiest part, with the rough side of sponge.

4. Wash clean with hot water from the tap, drain and dry.

· **貼心叮嚀**：洗鍋時，放一塊微濕的抹布在水槽檯面，可幫忙止滑和載重，洗起來較不費力；好的不鏽鋼鍋是耐用又安全的烹調道具，不會產生化學反應，故每次用完都刷洗乾淨，鍋具清潔又美觀，料理更得心應手。

· **Remarks**：Place a wet rag on the sink top for the whole cleaning process to position the pan rigidly and also to hold its weight. Stainless steel, besides its durability is the safest material for cooking, with no chemical reaction. For best cooking performance, appearance and hygiene, the pans have to be cleaned thoroughly after each use.

原味烹調的簡單基礎刀工

在此單元，不講究專業的刀工，而是教導如何保留食物原味的基礎刀工法及方便切菜的技巧，達到在烹煮過程中，鎖住食物營養到最高點。所以煮熟的時間不同，切長或切短的角度也不同。了解如何依食物的屬性作正確的切法，可以讓每種食材都能同步煮的剛好熟透，不僅吃到營養，也滿足享受美味的需求。

In this teaching unit, we do not stress on the professional cutting skills but rather on the simple basic cutting skills for maximum retention of natural nutrients. The main cutting theory is the understanding of the nature, its natural properties and the shape of each individual ingredient and cut them in their relative proportion so that they can be just cooked at the same time. In this manner, you will not only meet the approval of elderly folks who likes flavorful food, but you are in for a more nutritious and heavenly fare.

切條（瓠瓜）┃ Cutting Stripes (Bottle Gourd)

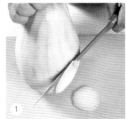

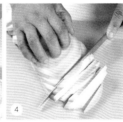

1. 去除外皮要薄，避免流失營養；先直立切下二片再橫倒就不易滑動而切傷手指。

2. 整齊切成1公分厚的片狀。

3. 保留瓜梗較好握住，並切到尾端。

4. 將所有的片狀排放整齊，才可以有效地將全部切成1公分的條狀。

1. Peel thinly for more nutrients, vertically cut two slices so it can lie flat horizontally and will not rock, to avoid the risk of hurting your fingers.

2. Cut into 1 cm even slices in a neat manner.

3. Reserve part of the stem to hold at the end.

4. Cut slices into 1 cm stripes, evenly. Make a neat job of it to speed up and facilitate cutting.

切絲（紅蘿蔔） ▎ Shredding （Carrots）

1. 帶皮保留皮下的營養素再直接切成斜薄片。

2. 依料理需求切成均等細絲。

1. Need not peel as peeling removes vitamins and mineral directly beneath the skin, cut neatly into thin even diagonal slices.

2. Then cut into even shreds, the thickness vary in accordance with the matching ingredients.

切斜片（小黃瓜） ▎ Cut Diagonal Slice （Cucumber）

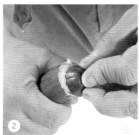

1. 帶皮直接切除兩邊的蒂頭。

2. 切斷的蒂頭放在切口處，左右來回磨擦兩頭，以去除苦味。

3. 再斜切成均等片狀。

1. Do not peel, trim both ends.

2. Place the end piece back on the cucumber, rub back and forth to eliminate the bitter taste on both ends.

3. Slice it diagonally and evenly.

切斜片（紅甜椒） Cut Diagonal Slice （Red Bell Pepper）

1. 先目測寬度約2.5公分。
2. 用細長的小刀從蒂頭均勻往下切開。
3. 去除內部的種籽。
4. 再切成2.5公分寬的斜片。

1. Cut about 2.5cm wide slice segment.
2. Cut down from the stem to the bottom evenly, with a paring knife.
3. Clean out the seeds.
4. Slice it diagonally, 2.5 cm wide.

· **貼心叮嚀**：甜椒要選外觀平滑，果肉厚實者佳。
· **Remarks**：Select a bell pepper with a thick meat and in good shape.

切丁（洋蔥） How to Dice （Onions）

1. 去除外皮及頭尾再剖半。
2. 用手壓住洋蔥再分層均等橫切至底部但不切斷。
3. 切至頂端再轉90度方向。
4. 全部均等切成小丁，到底時要稍微用力切斷。

1. Peel, wash, trim ends and cut in half.
2. Hold onto the top and cut even slices horizontally, leaving one end intact.
3. Cut from wide base to top and change direction and turn 90 degrees.
4. Then dice evenly, slightly chop when reaching the end that is kept intact.

切段（蘆筍） How to Cut in Segments (Asparagus)

1. 頭部較窄先直切90度，第二段切60度。

1. Cut the head at a 90° angle due to its narrowness, then cut a 60° angle for the second cut.

2. 第三段則切45度。

2. Cut at 45° for the third cut.

3. 底部較寬厚切30度。

3. Cut at 30° for the last cut due to its wideness.

4. 蘆筍依直徑寬窄的烹煮時間不同切成不同角度，使能同步煮到剛好熟透。

4. Cutting an asparagus at different angles in accordance with the width, to enable all segments to be just cooked at the same cooking time.

西洋芹切法 Cutting Celery

1. 每根自莖幹底部用刀撕除表皮較老的莖幹纖維。

1. Rib off the tough fiber at a stretch from the outside stem from the stalk end with a knife.

2. 刀片以45度橫切成較寬的斜片。

2. Slant knife to cut diagonally at a 45° angle for a wider surface cut.

切圓塊狀（白蘿蔔） ▎ Cut Cylinder-Shaped Block (White Radish)

1. 去皮並用食指測量厚度。

1. Peel, measure with your finger to cut into the right proportion.

2. 用刀子往下直切成圓塊狀。

2. Then cut into cylinder-shaped block.

切末（蒜頭） ▎ How to Mince (Garlic)

1. 蒜頭平坦處貼放砧板上，全部先切成均等薄片。

1. Cut neatly into thin even slices by placing the flatter side on the board.

2. 每次取少量片狀切成均等細絲。

2. Slightly stack a few slices and cut into stripes.

3. 最後切成均等細末。

3. Then mince evenly.

· **貼心叮嚀**：以此手法切成的蒜末比傳統拍碎再重剁的蒜末較乾爽輕巧，只需很少的油就可爆香又蒜味十足。

· **Remarks**：Manually mince garlic is preferred to using a chopper because it is even, light and more fragrant using only minimum oil.

切朵狀（綠花椰） ▌ Cut Flowerets（Broccoli）

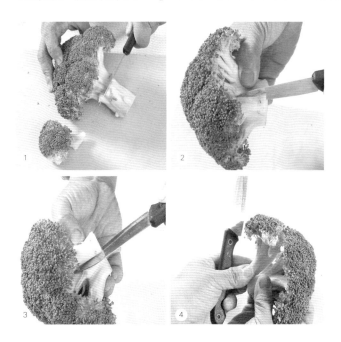

1. 距離花朵2公分處切斷過長的莖幹。
2. 用細長的小刀自花朵右側直切至莖幹尾端。
3. 再用小刀自花朵左側直切至莖幹尾端。
4. 用手取下並依序將全部切成小花朵後，依花朵大小稍微調整莖幹，才能同步熟透。

1. Trim off the stem, leaving about 2 cm of the stem intact.
2. Cut the stem from the right side of the floret all the way to the stem end, using a paring knife.
3. Then cut the stem from the left side of the floret all the way to the stem end.
4. Pull away one floret and repeat, then cut the part of the stem that has been trimmed off at the beginning in proportion to that of the floret.

· **貼心叮嚀**：如果花朵較小，可以2～3朵合併切成一大朵；不要從花頂切割，避免花屑掉落太多。
· **Remarks**：If some florets are smaller, combine 2 or 3 to make a floweret; to prevent crumbs of floret creating a mess, avoid cutting from the top of the floret.

向食物借油借水的烹調原理

向食物借油的烹調原理，是利用動物本身的脂肪烹煮；**向食物借水**的烹調原理，則是利用食物本身的水分，或添加少許水即可煮熟食物。

無油無水烹調的形成在於其烹煮過程中，沿著鍋蓋產生的蒸氣水膜，加上從鍋底到鍋身立體傳導的熱循環，讓食物在半真空的密閉空間，利用其本身水分所產生的蒸氣就可以完全烹煮到剛好熟透，讓食物健康又美味。

無油無水烹調要領的簡單基本要求：

依食物的量選擇尺吋或容量合宜的不鏽鋼鍋具。

烹煮過程不要一直開鍋蓋。開蓋煮食無法形成蒸氣水膜，熱氣和水分就容易流失，烹煮時間也會加長。

養成以水珠測鍋溫的習慣，並使用正確的溫度如中大火、中火、中小火及小火來控制烹煮過程的火候。

The principle of "**greaseless cooking**" means to cook with the fat that is naturally found in the meat; "**waterless cooking**" is to cook foods in their own natural moisture, or adding a little amount of water when needed. It is possible because a vapor seal is created around the lid, and heat is distributed evenly across the bottom and up the sides of the pan, to allow foods to be cooked completely in a bath of steam from its own natural moisture.

The simple basic rules are:

Always use the right size pan, one that the food is almost full.

Resist the urge to peek; when the lid is uncovered during the cooking period, the vapor seal is broken and the heat and moisture are allowed to escape, thus lengthening the cooking time.

Testing for the correct pan heat with water droplets and controlling the right heat use from medium-high, medium, medium-low, to low during cooking process.

鮭魚 ▌ Salmon

1. 依量取剛好大小的平底鍋，預熱至水珠快跑後，放入鮭魚肚，皮朝下輕壓使其完全貼平鍋底並逼出油脂。

1. Preheat an appropriate size skillet till water droplets "dance quickly"; place the salmon belly skin side down in pan lightly pressing down to ensure skin surface is entirely in contact with pan and traces of fat is being forced out.

2. 加入適量的鹽，加蓋以中火乾煎鮭魚（約2.5公分厚）約1.5分鐘，轉中小火繼續乾煎2分鐘。

2. Sprinkle a dash of salt, pan-broil salmon（2.5 cm thick）for 1.5 mins on medium heat, then medium low for 2 mins, with the lid on.

3. 轉小火後開蓋，用濕布擦鍋緣，翻面以中小火繼續乾煎2分鐘，由於沒有魚皮保護，不要煎過久導致魚肉乾澀。

3. Turn to low heat, uncover, clean the edge of pan with wet cloth, flip over; then pan-broil on medium low heat for 2 mins until desired doneness; as there is no skin to protect on this side, do not fry over 2 mins or it will become dry.

4. 熄火用濕布擦鍋蓋後即裝盤並馬上食用。

4. Turn off heat, wipe lid with wet cloth, serve immediately.

· **貼心叮嚀**：若是乾煎帶皮的整條魚，第一面以中火乾煎約3分鐘，第二面以中小火乾煎約3分鐘，若魚身較薄（低於3公分厚），第二面則以中火乾煎約2分鐘即可；若魚身瘦長，可自腹部切成二半放入鍋中，可避免鍋子太空而燒焦。

· **Remarks**：To pan-broil whole fish with skin, first side for 3 mins over medium fire, then second side for 3 mins over medium small fire. If fish is thin (less than 3 cm), second side for 1.5 mins over medium fire; for slim fish, halve it to nearly fill the pan so it will not cause the fish to burn.

雞腿肉 | Chicken Drumstick

1. 依量取適當的平底煎鍋，預熱至水珠快跑狀，放入雞腿並將皮壓緊平貼鍋底（兩面都醃鹽並將肉朝上），撒適量黑胡椒，加蓋。

2. 計時以中火乾煎約4分鐘至水氣冒出，轉中小火繼續煎2～3分鐘至熟。

3. 轉小火開蓋用濕布擦鍋緣，再用剪刀檢測雞腿最厚處，雞肉容易剪開又呈粉紅多汁鮮嫩感，即成。

4. 檢視雞皮是否酥脆金黃，若不夠酥脆，繼續以中大火乾煎約30秒至1分鐘，起鍋食用時，記得將脆皮面向上。

1. Preheat an appropriate skillet till water droplets "dance quickly". Place the chicken drumstick (both sides seasoned with salt) meat-side up firmly in the skillet, pressing down to ensure skin surface is entirely in contact with the pan and traces of fat is being forced out; sprinkle black pepper, cover.

2. Set the timer and pan-broil over medium heat for about 4 mins until steam appears, then turn to medium-low heat for another 2 to 3 mins until done.

3. Turn to low heat, uncover, clean the edge of pan with wet cloth; check for doneness by cutting the thickest part of the chicken with scissors; it should be easily cut, pinkish in color, tender and juicy.

4. Get a peek at the skin to check if it is crispy brown, it not, fry for another 1/2 to 1 min over medium high heat; serve with the crispy skin side up.

明蝦 | King Prawns

1. 明蝦先用紙巾擦乾水分後，撒黑胡椒於蝦腹，粗鹽置於鍋底，頭尾交叉站立放入適當的平鍋底（最好是用電烤鍋並預熱至400°F）。

2. 計時加蓋烤約6分鐘至熟（蝦子重約120克）；若不是用電烤鍋，先以中火乾烤約3分鐘至小水氣冒出，轉小火繼續乾烤約3分鐘至熟。

3. 打開鍋蓋完成，起鍋前，用不鏽鋼製食物刺針檢測靠近蝦頭的蝦腹，容易刺穿，感覺有彈性且外殼沒有變白，蝦身沒有捲縮狀，即可馬上食用。（若以鐵板電烤鍋預熱至400°F恆溫乾烤效果更佳。）

1. Pat dry, sprinkle black pepper on the stomach, and sea salt evenly in the skillet; place them in an alternate order of heads to tails and lay them on their stomachs preferably in an electric griller preheat at 400°F.

2. Set the timer for 6 mins (120g per prawn) with lid on until done; for non-electric skillet, pan-grille for 3 mins over medium heat until little steam appears and turn to low heat for about 3 mins until done.

3. Check for doneness with a cooking pin by piercing through the thickest part of the stomach, near the head; it should be easily pricked, have an elastic texture, the shell should not turn white, the meat should not shrink and the prawn should not curl up if it is not overcooked; serve immediately. (Best on griller at constant heat on 400°F for an elastic texture and to emerge its fragrance).

可以向哪些食物借水烹調圖解示範

葉菜類：菠菜 ▌ Leafy Vegetables: Spinach

1. 菠菜鬆開擺入冷的鍋中（勿壓扁並盡量鋪滿），入適量過濾水潤濕後再濾乾。

2. 均勻撒入適量的鹽，加蓋。

3. 計時轉中火煮約1分鐘至小水氣冒出。

4. 打開前用濕布擦鍋蓋，稍微翻轉菠菜使均勻，並馬上取出直接切食以避免鍋子的餘熱將食物煮過熟。

1. Place spinach loosely in cold skillet, one that most nearly fills without pressing it down; rinse in filtered water, then drain.

2. Sprinkle a dash of salt evenly, cover.

3. Set timer and pan-boil on medium heat for 1 min, till little steam appears.

4. Wipe lid with wet cloth, overturn the spinach to cook thoroughly; remove immediately and cut; the retained heat in the pan will overcook the vegetables.

根莖花果類：綜合蔬菜鍋 | Rhizome, Gourd & Fruit: Mix vegetables

1. 蔬菜梗交錯疊放鍋底。

2. 依蔬菜熟度如綠花菜梗、胡蘿蔔、白蘑菇、柳松菇等往上排放，快熟的如白菜葉交錯疊放在最上面，撒鹽調味再依量加水或雞湯或味醂至鍋子的半公分高。

3. 加蓋計時以中火煮約3～4分鐘至大量水氣冒出（白蘑菇不需太大，花菜及胡蘿蔔也不要切太大，以避免與其它的菜類煮時間差異過多）。

4. 打開鍋蓋前記得先用濕抹布擦鍋蓋讓營養回流，熄火後馬上食用。

1. Spread and stagger in circles and layers the stems at the bottom of the skillet.

2. The kinds of vegetables that take longer time to cook are placed at the bottom such as broccoli stem, carrot, white mushroom; the ones that cook quicker are placed on top such as the Japanese buna shimeji mushrooms and white cabbage leaves (lay out the same as stems). Sprinkle salt first, then add 1/2 cm of water or chicken stock or mirin in the pan.

3. Set the timer for about 3 to 4 mins over medium heat till much steam appears. (The selection of mushroom, broccoli and carrot should not be cut too big to avoid the various vegetables' cooking time to be so different.)

4. Remember to wipe the lid with a wet cloth to condense the nutrition before opening it. Serve immediately.

蛤蜊 | Clams

1. 先將蛤蜊擺入適當鍋中。

2. 利用空隙處加入蔬菜配料。

3. 加蓋計時，視蛤蜊大小以中
   火煮約3～4分鐘。

4. 水氣冒出即熄火，若蛤蜊超
   過120克則轉小火繼續煮約
   30秒～1分鐘，至蛤蜊微開
   再熄火，開鍋蓋用濕布擦鍋
   沿並馬上取用。

1. Place clams in the correct size skillet.

2. Add vegetables to the vacant spaces.

3. Set the timer for 3 or 4 mins
   over medium heat, depending on the size
   of the clams, with the lid on.

4. When steam appears, turn to low heat (for
   big clams above 120g) for 1/2 to 1
   min until all clams are slightly opened,
   and turn off the heat; wipe lid and edge
   of pan with a wet wipe lid and edge
   immediately.

Upgrading fine food is easy with just a little bit of creative power and skills. In the immediate future, you will be able to open your vision for fine food with new impressions and enabling you to enjoy the appealing taste of the finest silver-service restaurants at its height, even at home.

Exploit Creativity for Variations

We can utilize the shape of the food itself as decoration, like cross two pan-broiled lamb chops or stack them up in a different style; utilize another dish such as "Teppanyaki Eggplant Topping" (see page 162) or seasoned fruit such as mango with lime juice and cut as you like for garnish. As for "Pumpkin Boat", you can choose an extraordinary plate to serve an exquisite amount per serving, creating this like a naturally art object so tantalizing and extremely amusing.

Variance with Food Plating

"Sesame Chicken Salad" (see page 116) can be served in a particular platter with various divisions. Serving on the pan without plating from the stove directly to table, especially for rice dishes during parties will also make the dish looks palatable and dainty.

Variance with Different Utensils

It is easy to give shape to salad and fried rice, by utilizing a variety of pudding jelly molds. "Taro Pumpkin Pie" (see page 166) can be plated, using different sized moulds or a cocktail glass, press firmly and demould. As for "pumpkin paste", you can spread on a thin pancake and pan-fry until both sides are golden brown, creating a nutritious and tasty dessert. You can use a variety of bake moulds or ladle to make different lovable shapes or rounds for "Wheat Pancakes", and serve with various flavors of jams and maple syrup, making it an appetizing and irresistible 5-star tea time snack.

讓原味健康升級成五星級料理

其實美食升等只要發揮一點點的腦力及巧思，利用短短幾分鐘做個小改變，即能開啟美食視覺的新觀感，在家也能享受五星級美食的好滋味。

利用巧思做變化

利用食物本身的形狀做造型裝盤，例如乾煎兩塊羊排可互相交叉呈站立狀；再以「魚香茄子」（參考本書P.162）作配菜；或「檸檬芒果丁」點綴裝飾；或是南瓜船可選用合適的盤子分裝成一人份的精緻量，造型可愛又好吃。

利用餐具做變化

「涼拌芝麻雞絲」（參考本書P.116）可取用分隔容器裝盤；或是將全部的食材如「彩虹菜鍋」裝入一個合適的鍋具，直接上爐煮好或是一鍋熱熱的雞飯整鍋端上桌食用，看起來美味又可口。

利用器具做變化

沙拉或是炒飯可利用各式布丁果凍模具壓成可愛的型狀做變化。「芋頭南瓜派」（參考本書P.166）可分裝至各式大小不同的容器，壓成型再脫模；而南瓜泥可用春捲皮做成薄餅狀乾煎成南瓜薄餅，美味健康又好吃。製作全麥無酵鬆餅可善用各式可愛造型的燒烤模型；或用湯杓做大小圓型，再搭配不同的果醬、楓糖漿做成五星級下午茶點心。

073

PART 3
讓原味更豐富的
調味料、辛香料與佐醬

Spices and Sauces

烹飪如同繪畫，是一門藝術。

把各樣不同的食材加以組合變化為許多美味的料理。

天然食物的鮮味遠超過烹調五味──甜、酸、苦、辣、鹹；所以健康烹調不只是強調保留食物的營養價值和原味，也要能滿足我們的味蕾。

有時藉著不同味道的組合重整，可以增加甜度或產生新奇的口味變成更好吃的料理；而「愛」是最好的調味料！

用愛調味可以讓食物產生更大的魅力，使人吃了更享受。

Cooking is an art, just like painting.

You combine or mix various ingredients into many delicious dishes; the delicate flavor of the freshness of natural food is far beyond the five tastes which is sweet, sour, bitter, spicy and salty.

However, healthy cooking not only emphasizes the skills to preserve the nutritional value and natural taste, it also satisfies our taste.

Sometimes, a combination and the blending of different flavors can enhance the sweetness, give a harmonizing and superb taste, and make it a more appetizing and excellent dish.

Above all, LOVE is the most important spice; if you add love to any recipe, there is a great tendency that everyone will enjoy your food.

Cooking for original flavor using oil extracted from ingredients (meat), not only reduces the amount of oil you use, but it also forces out extra "saturated fatty acid", increasing the health value. The selection of cooking oil, therefore, is even more important. Consider whether the ingredient is good for salad, stir fry, or deep fry and then use quality thermostable oil.

Knowing Fat and Cholesterol

Fat is the main source of calories, providing important nutrients for your body. Naturally, fat comes from animals and plants, but only animal fat contains cholesterol. Cholesterol is a wax-like material that stays in the human body. Cholesterol can originate in two ways: produced in the liver or absorbed from foods such as eggs, meat, fish, dairy products, pork, oil, etc. However, cholesterol from food is not the main cause of having an unhealthy amount of cholesterol in the blood. Instead, the body softens saturated fatty acids into cholesterol, and then stores it in the liver and the blood. Therefore, it is important to decrease the intake of saturated fatty acids.

All fat contains three types of fatty acids: saturated fatty acid, poly unsaturated fatty acid, and mono unsaturated fatty acid. Intake of saturated fatty acid will increase cholesterol in the blood. Poly unsaturated fatty acid has been proven to lower cholesterol, but it is easily oxidized, producing free radicals. Mono unsaturated fatty acid, on the other hand, is the least likely to increase blood cholesterol and is not easily oxidized.

Choosing the Right Oil – The First Step to Healthy Cooking

Borrowing oil from the ingredients means to extract oil directly from the ingredients. How and what you cook will determine the kind of oil you use. When deep-frying or stir-frying, you should use thermostable vegetable oil like the Canadian canola oil. When you use oil in a salad, you can use sesame oil, olive oil and even flaxseed oil. It is recommended to regularly switch between a variety of oils since each have a different cooking effect and nutritional value.

讓原味加值的調味料

油

　　了解如何向食物（動物類）借油進行原味烹調，除了減少額外的用油量，也可逼出多餘的「飽和脂肪酸」，讓料理健康加分。因此在食用油的選擇上更加重要，應依料理的屬性如涼拌、清炒或油炸，選用穩定性佳的優質油品。

認識油脂及膽固醇

　　油脂是熱量的主要來源，提供很重要的養分。在自然界中，油脂來自動物及植物，但只有動物油脂含有膽固醇。膽固醇是一種存在人體內，軟如蠟般的物質，主要有兩個來源：由肝臟製造；由食物中攝取，如蛋、肉、魚、奶製品、奶油及豬油等。然而，食物中的膽固醇並不是造成血液中膽固醇濃度偏高的主要原因，而是身體利用飽和性油脂將之軟化成膽固醇，而堆積在肝臟與血液中。因此，減少飽和性油脂的攝取是非常重要的。

　　所有的油脂都包含三種脂肪酸：飽和脂肪酸、多元不飽和脂肪酸、單元不飽和脂肪酸。攝取飽和脂肪酸會提高血液中的膽固醇；多元不飽和脂肪酸已被證明可以降低血液中的膽固醇，但易被氧化產生自由基；而相對地單元不飽和脂肪酸最不易增高血膽固醇的濃度，也不易被氧化。

正確選擇油脂，原味健康料理的第一步

　　向食物借油是取之於食物本身的油；作菜時要炒要炸，就要依料理選擇不同的優質食用油，如油炸或快炒要用耐高溫、穩定性佳的植物油如加拿大芥花油；涼拌除了香油、橄欖油也可用亞麻子油，建議多重選擇，偶而替換，都有不同的烹調效果和營養價值。

Comparison of Dietary Fats

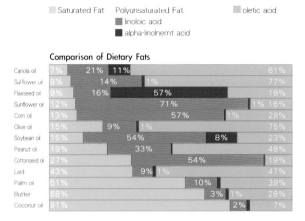

The greatest benefit of borrowing water from the ingredients is that it allows us to retain the highest amount of nutrient in our food and we can better present the original flavor in the food by just adding a little salt, sugar, vinegar or soy sauce. Even without using seasoning, you will enjoy the true natural tastes of foods. When we enhance the natural flavors of healthful food without adding excessive use of seasoning, we can then use seasoning based on the need of the ingredients without having to sacrifice the true natural tastes.

Seasons Using Natural Ingredients

Cooking for natural flavor not only retains the original flavor of the ingredients, it can also reduce the amount of seasoning you use. Many ingredients have natural flavors: the sweetness in onions and carrots; the saltiness in celery and tomatoes; the sourness in lemon and kumquats; and spiciness of pepper. All of these can be utilized in the natural flavor cooking or to decrease the amount of traditional seasoning you use.

Main Ingredients Contains Great Flavor   If the main ingredient is already full of flavor, only a small amount of natural seasoning is needed. Boneless short ribs, for example, when cooked with perfect timing, only need a small amount of black pepper to increase its fragrance.

Using Natural Vegetables for Flavor (ex. Onions, Tomatoes)   When using vegetables for seasoning, you do not have to rely on the use of chemical seasoning, such as monosodium glutamate(MSG), tenderizer, etc, but you can

■ 飽和脂肪酸　　多元不飽和脂肪酸：　　　　　　■ 單元不飽和脂肪酸
　　　　　　　　■ 亞麻油酸
　　　　　　　　■ 次亞麻油酸（Omega-3脂肪酸）

食譜中含脂肪量　脂肪酸含量規範化至100%　　　　　脂肪酸含量百分比

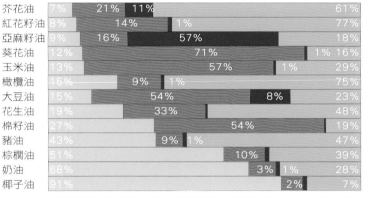

芥花油	7%	21% 11%	61%
紅花籽油	8%	14% 1%	77%
亞麻籽油	9%	16% 57%	18%
葵花油	12%	71%	1% 16%
玉米油	13%	57% 1%	29%
橄欖油	15%	9% 1%	75%
大豆油	15%	54% 8%	23%
花生油	19%	33%	48%
棉籽油	27%	54%	19%
豬油	43%	9% 1%	47%
棕櫚油	51%	10%	39%
奶油	68%	3% 1%	28%
椰子油	91%	2%	7%

來源：POS實驗植物股份有限公司，薩斯卡通，薩斯卡徹溫，1994年6月
（資料來源：加拿大駐台北辦事處提供）

利用天然蔬果就
能使原味升級又
健康，如檸檬的
酸味及香味。

　　向食物借水的烹調法，最大的魅力在於鎖住營養到最高點，真實呈現食材的原汁原味，只要一些鹽、糖，或一些醋和醬，甚至不需任何調味料就可以讓人吃得很享受。

　　以天然食材提味

　　天然原味的烹調法，除了保留食物的原味及營養之外，也能減少調味用量，因為在很多的食物裡面，本身已有天然的美味，如洋蔥、胡蘿蔔的甜；芹菜、番茄的鹹；檸檬、金桔的酸；辛香料的嗆辣等，都可運用在原味料理取代或降低傳統調味料的用量。

　　食物本身已有鮮美滋味：如食物本身已帶有濃厚的鮮美滋味，只要稍加天然調味料。例如無骨牛小排煎到剛好的熟度，只要撒入少許黑胡椒，即有增香提味的作用。

　　運用天然的蔬果提味（如洋蔥、番茄）：運用天然蔬果調味，不需化學調味料如味精、嫩精，就可以讓食物的味道升級又營養。如蘿蔔封肉，加了番茄和洋蔥，不需鹽、不需糖就讓原味更美味。

　　自製天然調味醬汁：將天然食材經適當比例調配成的醬汁，可以突顯主食材的魅力風味。例如蔥薑蒜酒汁可以幫肉類食物提鮮去腥；蒜片金桔醬可用於清煎魚肉的佐料；檸檬優格醬適合做為生菜沙拉的調味料；檸檬辣椒醬是肉類食物最佳佐醬，也可做為調味料使用。

079

將蔥、薑、蒜、米
酒各等量，放入果
汁機打成泥狀，擠
出成汁，即成蔥薑
蒜酒汁。

also improve the taste and nutritional value of the food. Wrapping the meat with radish and adding some tomatoes and onions can enhance the flavor of the dish without using salt and sugar.

Homemade Natural Seasoning Sauce   A proportional amount of natural spices as seasoning sauce can surface the authentic and captivating fresh flavors of the natural ingredients. The mixed juice from scallion, ginger, garlic and wine for example, can increase the fresh flavors while eliminating the meaty scent. The sliced garlic kumquat sauce may be used as a dipping sauce for fried fish; the lemon yogurt sauce is appropriate for salad dressing; the lemon chilly sauce is perfect as a dipping sauce for meats; and these sauces can be used as seasoning for other purposes as well.

Salt

Salt is an essential seasoning in cooking. It can also be used for preservation, sterilization and dehydration purposes. In recent years, low-sodium cooking has become popular. Many natural healthy salt products have appeared in the market. The traditional refined salt is saltier and is good for dehydrating vegetables, but it is also high in sodium content. Using different kind of salts for different kinds of cooking can create new sensations for your taste buds.

Bamboo Salt   A bamboo tube is first filled with natural seawater, and then it is heated repeatedly. After it cools, the crystallization becomes the salt. Through the baking process, the bamboo salt is rid of the heavy metals and it absorbs beneficial minerals from the bamboo. The bamboo salt tastes lighter than the common salt. It is suitable for making soup or seasoning your vegetables.

Sea Salt   Sea salt is made by drying seawater under the sun. Sea salt retains the precious elements from seawater and it is filled with flavor from the ocean. It is high in nutrients and it contains mineral substances that the body needs, such as iron, calcium and magnesium.

Mineral Salt or Rock Salt   Seawater, through the change of time,

常用的基礎調味料

鹽

烹調中不可或缺的調味料之一,同時具有防腐,殺菌,脫水的功能。近幾年,原味烹調興起,低鹽健康料理成為時尚,市場也出現許多天然的健康鹽品;傳統的高級精鹽,較鹹,適合脫水作醃漬品,但鈉含量也較高。針對不同的料理,選用不同的優質食用鹽,就可以激盪出不同的味蕾變化。

竹鹽

將天然日曬的海水填進竹筒內,經過反覆的加熱燒烤,冷卻凝固結晶提煉而成的。竹鹽經過燒烤可濾掉重金屬,吸收竹筒內對人體有益的礦物質及微量元素,鹹度比一般的鹽更清淡,適合用於煮湯或炒菜調味。

海鹽

是以海水做為原料,經過煎曬製成,蘊藏海水珍貴的元素,充滿天然海洋的風味,含有鐵、鈣、鎂等多種人體所需的礦物質,營養價值高。

海鹽

礦物鹽或岩鹽

來自於遠古海洋經過長時間的水分蒸發與地殼變動升起,歷經大自然的侵蝕,鹽分累積於地層逐漸結晶成岩鹽;鹽質較純淨、無污染,是最自然原始的風味,蘊藏豐富多元的礦物質精華比海鹽高約3～5倍,可廣泛應用在各種料理。色澤以純白色,質愈硬較佳,若是色澤偏於灰色表示雜質度較高。

evaporated, then the salt content accumulated and crystallized under the earth's crust, which then became rock salt. Rock salt is pure and unpolluted. It has nature's flavor and its rich mineral content is three to four times the amount in sea salt. It can be used in a variety of cooking. The whiter its lustre and the harder the rock salt, the better the quality of the salt. If the luster is greyish, then the salt contains more impurities.

Rose Salt Rose salt contains rich natural iron and the water-soluble calcium ion. When the lustre appears bright pink, it is the most precious kind of mineral salt. It can be used in cooking, in cosmetics, or as a health supplement. It is also very popular in Japan. It is often used in high-end Japanese Teppanyaki restaurants where they grind the rose salt into powder and then sprinkle directly over the food. It can add to the value of the dish with an elegant sweetness and will not affect the true natural essence of healthful food.

Coarse Salt Coarse salt comes from offshore seawater that is dried and crystallized under the sun. It is unprocessed sea salt and it has a rough texture. Coarse salt is good for softening dirt, cleaning and sterilization. It is also suitable for cleaning fish and meat.

Is Low Sodium Salt Healthy?

· Low sodium salt reduces the sodium content, but it does not mean lower salt intake. Instead, there is a higher content of potassium. Although some doctors will recommend to heart disease patients who need to decrease sodium intake to switch to low sodium salt, people with poor kidney functioning are forbidden to use it (low sodium salt and low salt soy sauce in the market usually substitute sodium with potassium ionic). Excessive potassium ions in your heart can affect heart functioning.

玫瑰鹽

含豐富天然鐵質及水溶性鈣離子，色澤呈自然光亮璀璨的粉紅色是極自然珍貴的礦鹽，其用途無論是使用於料理、美容、養生或食品等，在日本均獲得廣大的迴響。目前玫瑰鹽大多應用在日式鐵板高級料理，直接將玫瑰鹽塊研磨細末狀，輕巧地撒在食物上，可增加食材的價值感及趣味性，加上玫瑰鹽味道淡雅回甘，不會搶走高級食材的原味。

玫瑰鹽

粗鹽

取自近海的海水經日曝曬結晶形成，是未經過處理的海鹽，顆粒狀較粗糙。粗鹽有軟化污垢、清潔及殺菌的作用，很適合用來清洗魚肉類等食物。

粗鹽

如何聰明吃對低鈉鹽？

· 低鈉鹽是減少鈉含量，但不表示就是少鹽，反而增加鉀含量。雖然有些醫生會建議心臟病患者少攝取鈉而改用低鈉鹽，如果腎功能不佳者必須禁止使用（市售低鈉鹽、薄鹽醬油、無鹽醬油多半以鉀離子取代鈉離子），以免心臟堆積過多的鉀離子，影響心臟功能。

Sugar

Natural food also has natural sweetness. Greaseless and waterless cooking can reduce the use of sugar. Foods or desserts with high acidity may still need to be cooked with a well-selected sugar.

Red Sugar is also called black sugar. It is refined from sugar cane. The luster tends to be reddish brown and the taste has a hint of sugar cane aloe. It is suitable for increasing the fragrance of the food.

Crystal Sugar may be further categorized into golden crystal sugar, gram crystal sugar and gem crystal sugar. It does not turn moist or sour easily and it is sweet but not tiring. Golden crystal sugar comes from sugar cane that has undergone multiple crystallization refinements and the lustre tends to be golden yellow. Gram crystal sugar is a high-purity crystal sugar, mainly produced by melting and crystallizing white sugar. Golden crystal sugar is mainly suitable for black tea, coffee and other cold or hot drinks. Gram crystal sugar and gem crystal sugar are good for cooking braises, making jam or dessert soup.

Brown Sugar or cane sugar comes from dissolved sucrose that crystallizes after a filtration process. The brown colored sugar is suitable for desserts, pastries and drinks.

Icing Sugar is the result of mixing sugar cane and liquid sugar. Its lustre is pure white and it dissolves immediately in your mouth. It is suitable for dipping fruit, dumpling rice, pancakes and pastries.

Malt Sugar comes from starch that is dissolved in malt extract. It can be used for fruit jam or to increase the food's appearance and nutritional value.

Honey contains quality sugar and beneficial vitamins and minerals. Its fragrance can differ based on how it was harvested.

Maple Syrup being one of the most famous products of Canada, is a natural production with no coloring agents or additives and contains minerals and vitamins. It is the only alkaline sugar and it can take heat well. Its flavor is mellow and not as sweet as honey. Its smooth sweet aroma also adds nutritional value to our meals.

糖

天然的食物也有其自然的甜分，故善用無油無水的烹調法料理食物，可幫助降低食用糖的用量，但有酸度的料理或甜品之類，仍需依其烹調的需求選用不同的優質食用糖。

紅糖

又稱黑糖，主要以甘蔗原料提煉而成，色澤呈紅棕色，味道有甘蔗蜜香，適合增色提香。

冰糖

又可分為金冰糖、克晶冰糖及晶冰糖等。冰糖不易潮濕及變酸，味道甜而不膩。金冰糖是甘蔗砂糖經過反覆多次的結晶提煉而成，色澤呈金黃色；克晶冰糖及晶冰糖則是屬於高純度的晶形冰糖，主要由精煉的白糖經溶解與多次結晶提煉而成。金冰糖主要適合用於紅茶、咖啡及各式冷熱飲料；而克晶冰糖及晶冰糖則適合用於燒滷調味增亮、作果醬、煮甜湯等。

砂糖

主要利用溶解的蔗糖，過濾雜質結晶而成的棕色砂糖，適合甜點、西點、冰品或飲料。

細棉白糖

以甘蔗砂糖和液糖調製而成，其色澤潔白純淨，入口即化的粉末狀，適合沾食水果、棕子、精緻糕點。

麥芽糖

由澱粉經麥芽酵素水解形成，可應用在果醬及食物中。

蜂蜜

含優質的糖分及有益健康的維生素及礦物質，因養蜜方式不同而具特殊的天然花果香味。

楓糖漿

含豐富的礦物質及維他命C，完全沒有添加色素及加工過的天然產物，也是加拿大特產之一。其味道香醇，甜度比蜂蜜低，是唯一鹼性的糖類，耐燒煮，口感滑順，且有特殊楓香，料理食物可以健康加分。

楓糖漿

085

Vinegar

Vinegar is the first seasoning made by mankind and it adds taste to the basic flavors of natural food. In the early days, mainly white and black vinegar were used. Today there are many kinds in the market, which can be divided into three categories: brewed vinegar, synthesized vinegar and processed vinegar.

Brewed Vinegar is made from grains and fruit with added acetic acid bacteria and ethyl alcohol. The fermentation process produces the sour taste. After filtration, it becomes the natural brewed vinegar. The acidity of brewed vinegar is temperate and it has a mellow fragrance.

Synthesized Vinegar is also called chemistry vinegar. It is made from first adding water to edible glacial acetic acid and then adding amino acid organic acid, pigment, essence and seasoning. It can be mass produced and its flavor is more intense.

Processed Vinegar comes from mixing an exact proportion of brewed vinegar with diluted glacial acetic acid. It is also called half synthesized vinegar. Black vinegar is an example of this. The longer the vinegar was brewed, the more expensive it is. It has a strong but graceful fragrance and a natural, beautiful color.

Comparison of Various Vinegars and Their Uses

White Vinegar is made with rice, wheat, alcohol and other materials. The taste is sour and strong. It goes well with white meats, seafood, soups, sauces and salads. It can neutralize food that is too salty.

Rice Vinegar is made from a fermented mixture of glutinous rice and brown rice. It has the strong taste of synthesized vinegar but it has a pure fragrance and a smooth taste. It is good for making salads and to preserve the color of vegetables. It can also be used for dipping, marinating or to make sushi rice.

Black Vinegar it's main ingredient is brewed vinegar with added spices, vegetables, mushrooms, seaweed, sugar, salt and more. It has a unique fragrance and colorful lustre. It is mainly used as seasoning in foods such as sweet and sour dishes, gluten soups, noodles or for matching sauces with strong flavor.

醋

醋是人類最初製造出的調味品，可讓原味料理有另種風味。早期主要是使用白醋及烏醋做為食物的調味料，如今醋的種類繁多，可概分為釀造醋、合成醋及加工醋等三種。

釀造醋

以穀類或水果做原料，加入醋酸菌及酒精發酵產生酸味，經過濾釀製成，是純天然發酵的釀造醋，酸度較溫和，香氣較香醇。

合成醋

又稱為化學醋，是以食用級的冰醋酸加水稀釋，再加上胺基酸、有機酸、色素、香精及調味料等材料，加速催化發酵製成，可大量快速生產，味道較刺激。

加工醋

以純天然的釀造醋及稀釋的冰醋酸，以一定的比例混合調製而成，又稱半合成醋，烏醋即是屬此類的加工醋。純天然的優質釀造醋，釀造的年分愈久價格愈高，且有優雅撲鼻的香味，呈自然漂亮的色澤。

各種醋品特色與用途比較

白醋

以米、麥及酒等材料釀造而成，味道酸嗆，適用白肉、海鮮、醃漬、湯品、沾醬或涼拌類，可中和味道太鹹的菜餚。

糯米醋

用糯米、糙米以一定比例自然發酵釀製而成，有化學醋的辛嗆，氣味香純，口感溫順，適用涼拌菜，可保持蔬菜色澤鮮亮；作沾醬、醃漬食物的底醋、料理或壽司飯皆宜。

烏醋

以釀造醋為基本材料，添加辛香料、蔬果、香菇、海帶、糖、鹽等材料製成，具有一股獨特香味，色澤豐潤，大多用於料理完成品的提味，例如糖醋料理、羹類、麵食或搭配口味重的醬料。

Red Vinegar is made from fermented red glutinous rice. It has a red lustre and its taste is gentle. It is good for emerging the fragrance of red meat and gluten soup, ex. Zheng-Chiang vinegar.

Red Wine Vinegar comes from fermented grape juice. It is filled with the fragrance of red grapes and has a pleasant sourness. It is able to neutralize taste and is good for tenderizing red meat. In western cuisine, it is often used with olive oil as a salad dressing.

Fruit Vinegar comes from mixing brewed vinegar with fruit. It has the natural sourness from fruit and is sweet to the taste. It is suitable as a salad dressing, dipping or cooking and is good for bringing out the freshness of fruit and vegetables.

Soy Sauce

Soy sauce is commonly used in Chinese cuisine for seasoning and adding flavor and color to food. The priority for the choice of soy sauce should be natural and without preservatives because original flavor cooking means maximum retention of the natural essence and juices of natural food. The taste and the color of soy sauce should be mild and cooking time should be short. In this way, it can blend the soy sauce with the basic flavors without sacrificing the original taste. Each type of soy sauce is special and our choice should depend on our cooking needs and personal preferences.

Aging Soy Souce is made purely from soybean and wheat. The brewing time is four to six times longer than other kinds of soy sauce. It does not contain any chemical seasoning or preservatives. It has a ruddy lustre and a rich bean fragrance. The price tends to be expensive. It can be used in various type of cooking.

Glycol Soy Sauce is made from a traditional process of mixing soybean, wheat, salt and sugar. The flavor is not too sweet or too salty. It has a natural flavor that helps increase the appetite.

紅醋

以紅色糯米為材料經發酵製造而成，色澤紅潤，酸味柔和，如鎮江醋適用於勾芡、提味之紅肉類食物。

紅酒醋

用葡萄汁發酵釀造而成，帶有濃郁紅葡萄香氣，酸味宜人，具有提鮮去澀的作用，適用於紅肉，使肉質鮮嫩；在西餐料理，經常與橄欖油合成油醋汁，淋在生菜沙拉調味。

水果醋

以純天然釀造醋、水果為材料經長時間釀造而成，帶有水果自然酸香味，口感芳香甘美，適用沙拉類的醬汁材料，可增加蔬菜及水果的鮮香味，還有涼拌、沾醬或料理。

橘醋

醬油

醬油是中式烹調廣泛使用的調味料，主要用來提味和增色。原味烹調因保留食物較多的原汁原味，在醬油的選用上，以天然、無防腐劑為優先考量，醬色及醬味不要太重，烹煮時間不要過長，就可以在料理中融入醬油香氣，又不會蓋住食物原味。每種醬油的釀製方法各有巧妙，了解不同的醬油特色，可幫助我們依料理需求或各人喜好作適當的選購。

陳年醬油

主要成分以黃豆及小麥100％純原汁釀造而成，釀造時間較一般的醬油多4～6倍時間，無添加任何人工香料及防腐劑，色澤紅潤，充滿濃郁豆香，味道甘醇，但價格較貴，可隨意運用在各式料理烹調。

甘醇醬油

主要成分以黃豆、小麥、鹽、糖等原料遵循古法釀造技術製成，味道不會過甜或太鹹，具有自然甘醇的醍醐味及開胃的吸引力。

醬油

Japanese Mushroom Soy Sauce is a quality brewed soy sauce. It is made from high-quality shiitake mushroom, bonito flakes or dried kelp (konbu), salt, sugar and many other ingredients. It does not contain preservatives and artificial color. It has a sweet taste with a natural flavor of soybean and bonito flakes. It is concentrated but not overly salty like other soy sauces. It is suitable for stewing, slow cooking, dipping, marinating and steaming. It can also be used in soup noodles and hotpot. Its uses are endless.

Low-Sodium Soy Sauce is made from soybean, wheat, sugar, table salt, ethyl alcohol and other ingredients. It is often made with a salt substitute to lower the sodium content. The soy sauce is suitable for patients who need to watch their sodium intake.

Kelp Soy Sauce comes from mixing kelp extract with naturally brewed soy sauce. Its flavor is fresh and pleasant. It is suitable for dips, soup noodles and soba.

"Tea Bottom Oil" is also known as the shade oil. It is made from first boiling then cooling black beans, which is then mixed with saccharomycetes. The beans turn white after seven days of fermentation under high temperature. Salt is then added to the mixture. After sitting under the sun for four months, the raw black bean juice sinks to the bottom of the pot and it is the first soy sauce made from pure fermentation, which is why it is named "tea bottom oil." It is the best among all kinds of soy sauce and it has a natural bean fragrance and a rich sweet flavor. It also has a high nutritional value due to the black bean content.

日本香菇醬油露

　　精選純釀造醬油，以高級香菇、柴魚或海帶、鹽及糖等原料，採傳統方式釀製而成，不含防腐劑及人工色素。品質優良，味道甘醇，帶有天然的黃豆清香及柴魚味，濃縮4倍卻不若一般醬油死鹹，適用醬燒、燉煮、沾、醃、蒸、麵食及火鍋湯頭等料理，堪稱萬能的醬色調味料。

薄鹽醬油

　　以黃豆、小麥、糖、食鹽、酒精、調味劑等原料製成的低鈉食品；大多使用代鹽成分來降低鈉的含量，也就是以氯化鉀取代氯化鈉，其鹽分含量低，適用飲食必須控制鹽分的患者。

日本香菇醬油露

昆布醬油

　　是萃取昆布的精華汁液，搭配純天然的釀造醬油製作而成，味道清爽鮮美不油膩，適用沾醬、麵類料理的湯頭或日式冷麵的沾醬等。

壺底油

　　即俗稱的蔭油。將黑豆蒸煮放涼，拌入酵母菌，經過7天高溫發酵至黑豆內部完全滲透呈白色狀，再拌入適量的鹽放入醬缸中，日曬4個月後，黑豆釀造的原汁會沉澱在缸底，是第一道純的發酵醬油，即稱為「壺底油」，更是醬油中的特級品，散發一股天然的豆香味，味道甘醇香濃，又有黑豆滋養功效，營養價值高。

091

壺底油

One of the secrets to cook tasty food is to understand how to utilize fresh herbs and spices. Each kind of herbs has its own flavor and fragrance, adding new layers of taste to your cooking. There are many kinds of fresh or dried herbs in the market. Most herbs come from fruits, flowers, leaves, roots, bark and other parts of the plants. Their rich fragrances not only increase the natural sweetness and aroma but are also beneficial to your health.

Natural spices are the sources of creative cooking. When orange peel is added to red bean soup, for example, it not only neutralizes the sweetness, but it also adds a fruity flavor to the dessert.

Onion is a nutritious vegetable that is full of calcium, phosphorus, iron and vitamins. It is usually used for flavoring in western cuisine. In Taiwan, winter is the harvest season for onions. Taiwanese onions are spicier and sweeter than imported onions. To avoid tearing when cutting onions, you can first rinse the knife under water or soak the onions in cold water for about ten minutes. (Example on page 144)

Celery has a strong fragrance. It can be eaten cooked or raw. From the color of the stem it can be categorized as green or white celery. The white celery smells like cilantro. The celery leaves are more nutritious than the stems. The leaves can be put in a salad or stir fried. (Example on page 146)

Ginger can be further categorized into aged ginger, powder ginger and young ginger. It has a unique spiciness and fragrance. It is full of vitamin A and C. Young ginger is good for pickles or eaten raw. Aged ginger tends to be spicier and it can warm our bodies. When ginger is added to food, it can eliminate the meaty and fishy scent in food. (Example on page 150)

讓原味加分的辛香料

　　把料理變好吃的祕訣之一，就是懂得善用各種天然的辛香料；因為辛香料具獨特風味和香氣，有多層次的豐富口感。新鮮或乾燥的種類眾多，大多取自於果實、花苞、葉、根莖或樹皮等部位，其濃郁的氣味，不僅能為食物增香提味，而且也蘊藏諸多對人體健康有輔助的療效。

　　天然的辛香料是創造食物巧妙的元素，例如紅豆湯加入新鮮的橘皮，除了解甜膩之外，又能增添果香味，令人回味無窮。

洋蔥

　　富含鈣、磷、鐵及維生素，是天然的健康蔬菜。西餐大多做為調味使用。台灣盛產期是在冬春之季，味道比進口洋蔥嗆味較夠，甜度較高。避免切洋蔥流眼淚，可先把菜刀用水沖洗，或是把洋蔥浸泡在冰水中約10分鐘。（示範食譜P.144奶香淡菜盅）

芹菜

　　有強烈香味，是香料蔬菜，可生吃也可熟食。從莖部的色澤可分為青芹及白芹等，白芹較細小，香味類似芫荽。芹菜葉比莖更營養，摘下的芹菜葉可涼拌、快炒，可吸收更多養分。（示範食譜P.146西芹番茄炒墨魚）

薑

　　分為老薑、粉薑及嫩薑，都具有獨特辛辣味和香氣，含豐富維生素A及C。嫩薑可醃漬成泡菜或生食，老薑的辛辣味道較重，烹調入菜可暖身去寒還可去除食物的腥味。（示範食譜P.150鍋烤龍蝦）

Garlic is the bulb of the plant. Its unique fragrance is spicy and stimulating. However, the strong taste is reduced when heated. It is indispensible in Chinese cuisine. It can sterilize meat and eliminate the meaty scent. Garlic is suitable for dipping or cooking in Chinese or western cuisine. (Example on page 170)

Hot Pepper or chilly is a pungent vegetable and pod pepper is the spiciest. Pepper seeds are spicy but turn sour under high heat. The pepper skin is mild. Its unique spiciness comes from its capsaicin content. It can resist oxidation, stimulate appetite, promote metabolism, sterilize food, eliminate unpleasant scent and help people stay warm. (Example on page 122)

Bead Onion is smaller than the green onion or scallion. The leaves are slender and the bulb is thick. Its fragrance is rich and full. It is also known as the "Big End Onion." It has a unique spiciness. The bulb is usually used to make dry crisp onions or shallots. The bead onion can be chopped or sprinkled in soups or stews. It is good as a seasoning for flavor or for a quick fry. (Example on page 204)

Basil has a rich but fresh fragrance. It is often seen in Chinese dishes such as three-spice chicken and fried clams. Basil is also used in western dishes such as pasta, pizza, seafood salad and pesto. (Example on page 162)

Cedar or Chinese toon has a strong fragrance. This vegetarian's favorite has the highest protein content of all vegetables. It is not suitable for stewing and deep frying especially under high heat; over cooking will make it bitter and sour. The tender leaves can be used for making various cedar sauces. It is good as seasoning for pasta, fried rice or meats, etc. (Example on page 124)

大蒜

　　是植物的鱗莖，其獨特氣味既嗆辣又刺激，經過加熱後減弱，是中式料理不可或缺的調味角色，有消除肉腥味及殺菌的作用，適合用於中西式的各種料理及沾醬等。（示範食譜P.170脆炒皇宮苗）

辣椒

　　是辛香味的蔬菜類植物，以朝天椒最辣。辣椒的種子很辣但不耐高溫會產生苦味，外皮的辣度中等。其獨特的麻辣感來自於辣椒素的成分，具抗氧化、刺激食慾、促進新陳代謝、禦寒、殺菌及去除腥味的作用。（示範食譜P.122和風銀芽肉絲）

珠蔥

　　其外觀體型比一般青蔥更矮小，葉細長，蔥白根處肥大，香味濃厚，又稱為「大頭蔥」，具特殊辛香辣味。其鱗莖大多用來製作紅蔥酥。可切碎或切末撒在湯或燉品，增香提味，或做為快炒爆香佐料。（示範食譜P.204原味什錦炒米粉）

九層塔

　　又稱為羅勒，其氣味濃郁清新。最常出現在中式料理的三杯雞、炒蛤蜊、西式料理的義大利麵、披薩、海鮮沙拉及青醬等食物，也會應用羅勒增添美味。（示範食譜P.162魚香茄子）

095

香椿

　　全株具濃烈氣味，其蛋白質為蔬菜之冠，深受素食者喜愛。不適合長時間燉煮，或高溫炒炸烹調易變酸或變苦；其鮮嫩的葉芽可調理成各式香椿醬做調味料用於義大利麵、炒飯、或肉排等。（示範食譜P.124香椿醬烤羊小排）

Coriander or Chinese parsley is often used in Chinese cuisine because of its intense fragrance. The leaves are easily over-cooked, so they are usually used as garnish. The stems, however, have a richer aroma and they are suitable for cooking soup, stews and quick frying. (Example on page 126)

Rosemary is a typical Italian houseplant. It is usually soaked in oil as a preparation before use. It has a rich fragrance but it is bitter and pungent. When added to meat, it can bring a fresh new fragrance and taste. It is perfect with lamb as it can eliminate the gaminess of the meat. (Example on page 118)

Mint is easy to grow. The leaves and stems are mainly used for eating. Its flavor is cool and fresh. It is suitable for different cuisine and desserts. It can also be used as a garnish. (Example on page 166)

English Parsley tastes cool but bitter. It brings a strong fragrance to the mouth and can be used as a breath freshener. It can be used with most ingredients or as a garnish. In western cuisine, it is often sprinkled on the food. It is suitable for soups, vegetables and sauces. The stem can be used for stews. (Example on page 148)

Bay Leaf has a sweet flavor and a rich fragrance. Just a single leaf can eliminate the meaty scent. It needs to be removed after cooking. Bay leaf, when overcooked, will turn bitter. It is suitable for Chinese or western meat dishes or soups. (Example on page 190)

Lemon Grass has a lemon taste and its aroma last longer than lemon. The stems not only can increase appetite, natural freshness and aroma; it can also act as a dehumidifier. It is suitable for meat and fish soups, hot drinks or desserts. (Example on page 112)

香菜

又稱為芫荽，是中式料理最常使用的配料。香菜葉烹調容易軟爛，所以大多切碎後點綴裝飾，而香菜梗的氣味比香菜葉更濃郁，適合燉湯、快炒或紅燒使用。（示範食譜P.126原味牛肩肉骨茶湯）

迷迭香

義大利的家常植物，多半浸泡在橄欖油做成迷迭香油，以備烹調使用。香氣濃郁，味苦略辛辣，與肉類搭配可增添清新的香氣及風味，尤其羊肉有去腥提味的作用。（示範食譜P.118迷迭香嫩煎梅花豬排）

薄荷

生長繁殖力強，容易種植，主要食用部位為葉及莖。味道清香涼爽，適用各式料理、甜點、飲品。（示範食譜P.166芋頭南瓜派）

巴西里

葉片味道清涼苦澀，入嘴有強烈清香，可抑制口臭。可與大部分食物搭配，常見辦桌用於裝飾食物；而西式料理則是切碎撒入食物上。適用於湯、蔬菜、醬汁，其莖部適合用於燉煮食物。（示範食譜P.148橙汁軟絲）

月桂葉

其乾燥的葉片，香氣濃厚，味道甘醇，只要放入一片即有去腥作用，烹調使用後，記得要取出，因為久煮會釋放出苦味，破壞食物的原味，適用中西燉煮的肉類或湯品。（示範食譜P.190紅酒燉牛腩）

香茅

含有檸檬的香味，但比檸檬香味更持久，以莖入菜除了增香及增加食慾的清爽感外，還有去寒除濕功效；適合製作肉類、魚類的湯頭、茶飲及點心料理。（示範食譜P.112香茅燉雞）

Pine Leaves or Pandan leaves have a sharp, long shape, with a light taro fragrance. This south-east Asian fresh herb is often used for pastries, cakes or deserts. Its extract can add natural color and aroma. Fresh pine leaves can be used for cooking and it allows the rich fragrance of the ingredients to emerge. Dry ones can be used to make tea. It is beneficial for patients with high uric acid and also good for protecting the liver. (Example on page 188)

Galangal a south-east Asian fresh herb, which resembles the ginger has a gingery, peppery and a light sourness like a lemon. In Malaysia and Indonesia, it is often used in curry or braises to give a special pungent aroma.

Star Anise which has a sensitizing licorice taste with a strong pungent aroma, is a basic ingredient for the stew pack. It also goes well with meat, whether braising or stewing and can eliminate the unpleasant meaty scent. Because of its strong flavor, only use it sparingly and avoid bitterness from overcooking. (Example on page 190)

Chinese Cinnamon comes from the bark of cinnamon trees. The closer it is retrieved from the center of the tree, the better the quality. Its natural freshness and spicy content, is a necessary spice for braising meat and can eliminate the unpleasant meaty scent.

Black and White Pepper Powder which is mainly produced in India, is a necessary spice in any kitchen. While the black and white peppers are most common, there are also red and green peppers. The colors are the results of different baking processes. The black pepper powder is spicier than the white one. It is suitable for marinating food or cooking western soup, while white pepper powder is more for Chinese dishes or soups. (Example on page 142)

Szechwan Peppercorn is usually used to eliminate the unpleasant meaty scent. When fried with salt, it can enhance the rich fragrance of the food. Its spiciness is more pleasant than pepper salt. In traditional cooking, Szechwan peppercorn should be oil fried in low heat, otherwise it will burn and lose its fragrance and spiciness. However, in healthy cooking it can be pan-fried without oil under low heat to retain its optimum fragrance. (Example on page 132)

七葉蘭

　　葉子呈長條尖狀，含淡淡芋頭香氣，是南洋的香料植物，經常使用做成糕點的香味及擠汁作色。新鮮的七葉蘭可用來料理烹調，有增香提味作用，而乾燥的七葉蘭適用泡茶，針對尿酸過高與肝火旺有良好療效。（示範食譜P.188紅豆糯米甜湯）

南薑

　　南洋的辛香料，其外型與薑相似，帶有薑、胡椒的味道及少許檸檬的酸味。在馬來西亞及印尼，經常用來煮咖哩或紅燒食物提味。

八角

　　具刺激性的甘草味及濃郁的香氣，是滷包最基本的材料之一，不論紅燒、燉煮，有去腥提味作用，但其味道濃並不適合放太多或煮太久，會產生苦味。（示範食譜P.190紅酒燉牛腩）

桂皮

　　取自於肉桂樹的樹皮，愈接近樹幹中心的樹皮品質愈上等，味道既芳香又辛辣。是紅燒不可或缺的辛香料，可去除肉類的腥味。

黑胡椒和白胡椒粉

　　主要產地是印度，最常見是黑、白胡椒，此外還有紅、綠胡椒等二種，主要是因烘培的程度不同。黑胡椒的辣度比白胡椒的辣度強，適合醃漬入味、西式濃湯調味等料理，白胡椒則常用於中式料理或湯品。（示範食譜P.142椒鹽烤明蝦）

花椒粒

　　大多用來消除肉類的腥味。運用花椒粒加鹽以小火炒香，即成花椒鹽，有增味提香效果，但是辛香味比胡椒鹽更芳香。傳統爆香花椒粒習慣用油炒，但溫度過熱則容易焦黑，失去特殊香氣及麻辣味；反之以無油低溫爆香較能保留其香麻味。（示範食譜P.132椒麻黃魚）

讓原味加分的佐醬

番茄莎莎醬 | Tomato Salsa Sauce

最佳賞味期：3天　　保存時間：5天
保存方式：移入冰箱冷藏室保存。
適合佳餚：前菜或搭配生菜葉，也可乾拌麵食、麵包、餅乾。
變化應用：捲在牛肉片中或搭配牛排，可讓牛肉吃起來不油膩。

> Best in 3 days　　> Store for 5 days
> Preservation : store and chill in the fridge.
> Suitable delicacies : eaten directly as an appetizer, as a dressing for salad, noodles, bread, or crackers.
> Varying application : wrapped in sliced beef or supplement the pan grilled steak it eliminates the greasiness and helps digestion.

此道私房醬汁完全利用天然食物本身的風味調合而成，不搗成泥狀，保留番茄的新鮮甜美，再搭配橘醋，口感微酸中略帶甜味。

材料：番茄丁300克、檸檬汁1/4杯、去籽紅辣椒末15克、冷壓芥花籽油1茶匙、九層塔末1大匙、蒜泥1茶匙、橘醋（或水果醋）50 c.c.

作法：全部的材料放入乾淨的容器中，攪拌均勻，即成。

The flavor of this specially formulated sauce is blended with the harmonizing flavors of only natural ingredients. You can retain the natural freshness and sweetness of the tomatoes by not over stirring them until pasty. With a little orange vinegar, the flavor is slightly sour but sweet.

Ingredients : diced tomatoes 300g, lemon juice 1/4 cup, minced chilly (seeds discarded) 15g, cold-pressed canola oil 1 tsp, minced basil 1 Tbsp, garlic paste 1 tsp, orange vinegar (or fruit vinegar) 50c.c.

Method : Put all ingredients in a container. Mix well, done.

· 貼心叮嚀：檸檬汁與辣椒丁拌勻，浸泡約5分鐘後，再放入其它的材料，可去除辣椒的辛辣味。

· Tips : Soaking minced chilly in lemon juice prior to adding other ingredients for 5 mins can neutralize the spiciness.

檸檬優格醬 ▌ Lemon Yogurt Sauce

最佳賞味期：3天　　　**保存時間**：7天
保存方式：移入冰箱冷藏室保存。
適合佳餚：水果生菜沙拉、涼拌蘆筍、綠竹筍、拌鮪魚或其他青菜如秋葵、玉米筍等。
變化應用：可加入水果，打成水果優格奶昔，再冷凍成冰塊，打成冰沙。

> Best in 3 days　　　> Store for 7 days
> **Preservation** : store and chill in the fridge.
> **Suitable delicacies** : as a dip for fruit, asparagus, bamboo shoot, other vegetables such as lady finger or corn shoot salad; or tuna paste, etc.
> **Varying application** : blend with fruit for a fruity yogurt smoothie, then freeze and re-blend into frozen slurry.

優格是整腸健胃的超級食物，帶有清新奶香味；搭配最恰當的檸檬汁、番茄醬及新鮮芹菜為佐醬，口感溫潤滑順，清香微酸，不甜不膩。

材料：原味優格1/2杯、美乃滋2大匙、檸檬汁2大匙、番茄醬1茶匙、中國芹菜末50克
作法：美乃滋、檸檬汁、優格、番茄醬放入乾淨的容器中拌勻，再加入芹菜末，移入冰箱冷藏即成。

Yogurt is perfect for protecting the stomach and regulating bowels. It has a fresh dairy fragrance. It is a perfect blend with lemon juice, tomato ketchup and fresh celery to give a refreshing, smooth and slightly sour taste that is neither sweet nor overpowering.

Ingredients : plain yogurt 1/2 cup, mayonnaise 2 Tbsp, lemon juice 2 Tbsp, tomato ketchup 1 tsp, minced Chinese celery 50g
Method : Mix mayonnaise, lemon juice, tomato ketchup and yogurt in a container. Add minced celery. Chill in the fridge.

101

· **貼心叮嚀**：每次取出，以一次食用的分量為主；建議以小杯沾食，會比淋在沙拉上拌食的用量少。

· **Tips** : Remove only the measure you plan on eating; the measure is smaller by using it as a dip instead of a dressing.

茄汁蝦皮辣醬 | Spicy Tomato Dried Small Shrimp Sauce

最佳賞味期：14天　　保存時間：30天
保存方式：移入冰箱冷藏室保存。
適合佳餚：適用於無水或少水烹調之綠色蔬菜和清煎的魚肉。
變化應用：炒麵條時加蝦皮辣醬，增色又增味，如同吃馬來炒麵般香辣過癮。

> Best in 14 days　　> Store for 30 days
> **Preservation** : store and chill in the fridge.
> **Suitable delicacies** :cooking vegetables waterless or in minimum moisture; or simple fried fish.
> **Varying application** : as seasoning for fried noodles for added color and flavor, similar to the Malay fried noodle, creating a more satisfying and spicy dish.

以辣椒、蒜頭為主，另加清香的小蝦皮及番茄，形成黃金組合的醬汁；口感類似傳統的馬來辣醬卻因使用新鮮的番茄倍增美味，又減少蒜頭辛辣感。

材料：紅番茄300克、小蝦皮35克、蒜頭25克、紅蔥頭25克、去籽大紅辣椒150克、去籽小紅辣椒2根、油50c.c.

作法：

1. 紅番茄洗淨，加入過濾水50c.c.加蓋煮約3分鐘，取出脫皮。小蝦皮洗淨，放入已預熱至水珠慢跑的8吋鍋中，以小火烤約6分鐘至香酥狀。

2. 將全部材料放入果汁機中，攪打成泥狀，取出倒入容器中。再倒入50c.c的芥花油及作法2的材料於8吋鍋中，以中小火煮沸並不時攪動，熄火，待涼，再裝入乾淨的玻璃容器即成。

This is a golden combination of chilly and garlic as the main spices and then adding dried small shrimps and tomatoes. It tastes similar to a traditional Malay spicy sauce but with the addition of fresh tomatoes, makes it more flavorful and also help to alleviate the pungent garlic spiciness.

Ingredients: red tomatoes 300g, dried small shrimp 35g, garlic 25g, shallots 25g, big chilly (seeds discarded) 150g, 2 small chilly (seeds discarded), oil 50c.c.

Method:

1. Clean tomatoes, cover and cook in 50c.c. of filtered water for 3 mins. Remove and peel. Clean dried small shrimp. Preheat a 8-inch skillet till water droplets "dance slowly" on the bottom. Pan grille the shrimp over low heat for 6 mins until crisp.

2. Blend all ingredients into paste. Pour in 50c.c. of canola oil and blended ingredients into the 8-inch skillet. Bring to boil over medium low heat, stirring occasionally. Turn off heat and cool. Put in clean glass container.

· **貼心叮嚀**：此道醬料是要吃番茄的鮮味，不需要另外爆香蒜泥及紅蔥頭泥。

· **Tips :** Focus on freshness of tomatoes. No need for frying garlic or shallots.

香椿青醬 ▎ Cedar Pesto

最佳賞味期：7天　　　**保存時間**：14天
保存方式：放入密封玻璃罐，移入冰箱冷藏室保存。
適合佳餚：適用炒飯拌麵、義大利麵及烤肉排。
變化應用：可塗厚吐司一起烤，或放在乾煎的雞腿肉
上提味，變成脆皮香椿雞排。

> Best in 7 days　　　> Store for 14 days
> **Preservation** : store and chill in the fridge in sealed
　glass container.
> **Suitable delicacies** : fried rice, noodle salad, pasta and
　pan grilled steak.
> **Varying application** : toast with bread, as a sauce to
　pan grille for crispy cedar drumstick to intensify the
　taste.

傳統青醬是使用九層塔，我們則使用特殊芳香氣味的香椿，搭配營養價值高的
松子及加拿大優質的芥花油，整體氣味更清新。

材料：香椿葉末100克、蒜末30克、松子末30克、腰果末15克、芥花油1/2杯、鹽
　　　1/4茶匙

作法：取3/4公升的小調味鍋預熱至水珠慢跑，入芥花油及蒜末一起爆香至淺金黃色。
　　　續入松子末及腰果末，以中火快速拌炒均勻，再加入香椿葉末及鹽拌勻，熄
　　　火，稍微攪拌即取出，待冷卻再裝入玻璃罐。

Traditional pesto uses basil, but we use cedar for its special fragrance. Adding
nutritional pine nuts and high grade Canadian canola oil further enhance the
freshness.

Ingredients: minced cedar 100g, minced garlic 30g, minced pine nuts 30g, minced cashew
nuts 15g, canola oil 1/2 cup, salt 1/4 tsp
Method: Preheat a 3/4-quart saucepan till water droplets "dance slowly" on the bottom,
add canola oil, stir- fry minced garlic over low heat until light golden yellow. In sequence, stir
in minced pine and cashew nuts, then add in minced cedar and salt, give a quick stir evenly
over medium heat till steam comes out. Turn off heat and remove immediately to cool.

· **貼心叮嚀**：香椿嫩葉須洗淨，擦乾水分，去除老葉硬梗，直接切成細末或打成細
　末狀，以手工切碎會比用機器攪打更有香氣，不需加油攪拌並可減低油量爆香蒜
　末。松子可放入塑膠袋中，用玻璃瓶壓成碎末狀。

· **Tips** : clean and dry the cedar well, discard old leaves and main veins, chop manually.
　Hand chopped brings out more fragrance than blending in oil with all other ingredients
　in a blender. In this way, it enables you to fry the minced garlic separately in minimum
　grease first, and adding the cedar in the end to preserve the original flavor and color.
　Pine and cashew nuts can be put in a plastic bag and mashed with a glass container.

檸檬辣椒醬 ▎Lemon Chilly Sauce

最佳賞味期：30天	保存時間：180天

保存方式：以玻璃瓶密封，放入冰箱冷藏室保存。
適合佳餚：椒鹽烤明蝦、新加坡海南雞、脫脂豬排、水煮白切肉或乾煎松板肉。
變化應用：取150c.c.檸檬辣椒醬，加入白醋100c.c.、糖2大匙、地瓜粉水1/4杯，一起煮沸，即成「泰式酸辣醬」。

> Best in 30 days	> Store for 180 days

> **Preservation :**store and chill in the fridge in sealed glass container.

> **Suitable delicacies :**pan grilled peppery salt king prawns, Singapore Hainanese chicken, low-fat pork leg, boiled pork in minimum water sliced, and defatted pork roast.

> **Varying application :**boil 150c.c. of lemon chilly sauce with 100c.c. of white vinegar, 2 Tbsp of sugar, and 1/4 cup of tapioca flour water to make Thai spicy sour sauce.

傳統辣椒醬較油膩及辛辣，此道辣醬卻以同等的檸檬汁及白醋，加上微酸的番茄醬及淡淡的蒜香和薑味，配搭原味的肉品創造出清爽又多層次的口感。

材料：新鮮辣椒600克、嫩薑100克、蒜頭150克、檸檬汁200c.c.、白醋200c.c.、黃砂糖3/4杯、番茄醬3/4杯

作法：除了檸檬汁，所有的材料用調理機打成泥狀再放入鍋中，以中火煮沸，加入檸檬汁並不時攪動至煮沸冒泡泡後，冷卻，裝入密封的玻璃容器，並移入冰箱冷藏室保存即可。

Traditional chilly sauce is more greasy and spicy. This chilly sauce combines a balanced amount of lemon juice and white vinegar, with a slight amount of tomato sauce for a light sour taste and light garlic and ginger taste. It is a wonderful dip for meats, creating different levels of taste to the flavor.

Ingredients : fresh medium size chilly 600g, young ginger 100g, garlic 150g, lemon juice 200c.c., white vinegar 200c.c., cane sugar 3/4 cup, tomato ketchup 3/4 cup
Method : Grind all ingredients to a paste in a blender. Bring to a boil all ingredients except the lemon juice over medium fire and then add in the lemon and re-boil until it bubbles, stirring occasionally. Allow to cool and store.

· **貼心叮嚀：**辣椒、蒜頭及嫩薑以小杵臼搗成泥狀的味道較香；新鮮調製但未煮開的辣椒醬較有鮮味，須在一星期內食用完。

· **Tips :** grind all the ingredients using the mortar and pestle instead of a blender can enhance the flavors and also creates a better texture. Freshly mixed raw chilly sauce is tastier, but must be consumed within a week.

芝麻醬 ▎Sesame Sauce

最佳賞味期：7天　　**保存時間：**14天
保存方式：移入冰箱冷藏室保存。
適合佳餚：適用涼拌菜、燙青菜、涼麵、雞絲涼粉、乾煎肉排。
變化應用：加入適量蒜泥、青蔥末、辣椒末，適合當火鍋肉片或燒肉的沾醬。

> Best in 7 days　　> Store for 14 days
> **Preservation :**store and chill in the fridge.
> **Suitable delicacies :**salad, boiled vegetable, cold noodle, shredded chicken bean jelly, pan grilled steak.
> **Varying application :** adding garlic paste, minced scallion and minced chilly for a good dipping sauce for sliced meats in hot pot or barbecue.

此道私房醬汁，以無油乾炒的方式，將新鮮的白芝麻慢慢炒至香味溢出，金黃飽滿狀；不苦也不燥熱，與其他的調味料融合成比傳統芝麻醬較不油膩的香醇滋味。

材料：新鮮白芝麻150克、冷壓芥花籽油50c.c.、芝麻香油1/2茶匙、味醂150c.c.、香菇醬油2大匙、水果醋50c.c.

作法：取11吋平底煎鍋，預熱至水珠慢跑（或鐵板電烤鍋預熱至325°F），加蓋，以小火慢炒至金黃色（略出油狀），偶爾要開蓋拌炒。全部的食材放入果汁機，攪拌均勻呈泥狀，取出，裝入密封的玻璃容器即成。

This specially formulated sauce emerge the fragrance of the white sesame by pan grilling them without oil until golden yellow and plump, with signs of oil on the surface of sesame seeds. It is neither bitter nor dry. When mixed with other seasonings, it is more delightful with a sweet aroma and less greasy, compared to the traditional sesame sauce.

Ingredients: fresh white sesame 150g, cold-pressed canola oil 50c.c., sesame oil 1/2 tsp, mirin 150c.c., mushroom-flavored soy sauce 2 Tbsp, fruit vinegar 50c.c.

Method: Preheat a 11-inch skillet till water droplets "dance slowly" on the bottom (or pan grille in electric griller, preheat at 325°F). Pan-grill covered, over low heat until golden yellow and plump (with signs of oil), stirring occasionally. Blend all ingredients into paste in a blender. Store in sealed glass container.

· **貼心叮嚀：**白芝麻在清洗時，要置放在不鏽鋼的小濾網中，用過濾水沖洗乾淨，再濾乾水分或曬乾，以避免太潮濕；鍋具要選寬淺的平底鍋較好翻炒均勻。

· **Tips：** When cleaning white sesame, place them in a stainless steel sieve and rinse with filtered water before drip drying. Wide and shallow pans with a flat bottom are more suitable to stir fry evenly and using the appropriate size gives best results.

蒜片金桔醬 Garlic Kumquat Sauce

最佳賞味期：3天　　保存時間：5天
保存方式：移入密封盒，再放入冰箱冷藏室保存。
適合佳餚：清煎的魚肉、炒米粉或炒麵。
變化應用：可另加香菜末及嫩薑末作成鍋貼、水餃、
火鍋肉片的沾料等。

> Best in 3 days　　> Store for 5 days
> Preservation : store and chill in the fridge in sealed
  glass container.
> Suitable delicacies : simple fried fish, fried rice-flour
  noodle and fried noodle.
> Varying application : can add minced Chinese parsley
  and ginger as a dip for pot stickers, boiled dumplings,
  and sliced meats in hotpot.

金桔本身具有自然回甘，酸中帶甜的特色，搭配蒜片、辣椒的清香微辣感調配成沾醬，適用肉類解油膩感，又有金桔的清香味，使口感呈現多層次變化。

材料：蒜片150克、金桔汁50c.c.、去仔紅辣椒片100克、香菇醬油1/2杯

作法：取一個乾淨的玻璃罐，放入蒜片、金桔汁、紅辣椒片、香菇醬油，靜置約15分鐘至入味，裝入密封的玻璃容器即成。

Kumquat is naturally sweet. It has a special hint of sweetness in the sourness. Here it is mixed with garlic and chilly as a light spicy dip. It complements well with various meats to eliminate the greasiness. The hint of kumquat with a naturally sweet aroma creates varying levels of taste.

106

Ingredients: sliced garlic 150g, kumquat juice 50c.c., sliced medium-sized red chilly (seeds discarded) 100g, mushroom-flavored soy sauce 1/2 cup

Method: Marinate sliced garlic and medium-sized red chilly with kumquat juice and mushroom-flavored soy sauce in a glass bottle and sealed, ready for serving after 15 mins.

· 貼心叮嚀：蒜頭直切成薄片，口感較佳。

· Tips : Sliced garlic gives a better texture.

黑胡椒蘑菇醬 ░ Black Pepper Mushroom Sauce

最佳賞味期：3天　　　**保存時間**：7天
保存方式：放入密封玻璃罐，移入冰箱冷藏室保存。
適合佳餚：適用紅肉排類，如牛排、小羊排、豬排或
海鮮如螃蟹，花枝等。

> Best in 3 days　　　> Store for 7 days
> **Preservation** store and chill in the fridge, in sealed
   glass container.
> **Suitable delicacies :**red meat such as steak, lamb
   chop, pork chop or seafood such as crab and squid.

先以奶油爆香洋蔥增加甜分，再入黑胡椒增香提味，利用麵粉讓醬汁產生少許稠狀，最後放入新鮮蘑菇片，比起一般的黑椒蘑菇醬，多了鮮嫩甜美的味道。

材料：奶油1大匙、麵粉1大匙、洋蔥末約150克、黑胡椒粒2大匙、香菇醬油2.5大
　　　匙、雞高湯1杯、蘑菇片150克

作法：

1. 取1公升調味鍋預熱至水珠慢跑，放入奶油以小火煮至融化，再入洋蔥末，加蓋，以
   中小火爆香1分鐘。轉小火，開蓋快速拌炒再入黑胡椒粒，加蓋，以中小火半分鐘，
   轉小火繼續與洋蔥一起拌炒至香及淺金黃色；入麵粉攪拌均勻。
2. 入醬油拌炒再倒入雞高湯，以中小火煮沸，轉小火，加入蘑菇片，以中小火再煮
   沸，轉小火燉煮約3分鐘，熄火即成。

First use butter to fry the onions to enhance the natural sweetness; and then add coarse black pepper to further intensify its taste and flour for thickening the sauce. Button-mushroom is added last for a more succulent and delicate flavor.

Ingredients: butter 1 Tbsp, flour 1 Tbsp, minced onion 150g, coarse black pepper 2 Tbsp, mushroom-flavored soy sauce 2.5 Tbsp, chicken broth 1cup, sliced button-mushroom 150g

Method:

1. Preheat a 1-quart saucepan till water droplets "dance slowly" on the bottom. Melt
   butter over low heat, add minced onions, cover for 1 minute over medium low heat.
   Turn to low heat, do a quick stir- fry, add black pepper, cover for 1/2 min over medium
   low heat. Turn to low heat, stir fry onions till fragrant and light golden yellow, then stir in
   the then stir in the flour evenly.
2. Stir in soy sauce. Pour in chicken broth and bring to boil over medium low heat. Turn to
   low heat, add mushroom, re-boil over medium low heat, then simmer over low heat for 3
   mins. Remove from heat, done.

· **貼心叮嚀**：在家烹煮少量醬料時，盡量用小調味鍋，聚熱快又均勻，可減少油或
　水的份量，讓醬料更香濃美味。

· **Tips：** when only cooking small amount of homemade sauces, use a smaller
   saucepan to reduce the need for oil or water and to make the sauce richer and more
   fragrant.

PART 4
讓原味更美味的**50道食譜**

原味烹調不單只是生食或水煮不加油、鹽、糖等調味料，而是經過正確的烹調過程，烹煮出比傳統更健康、比原味更美味的料理。它的最大特色就是「鮮」，而「鮮味」口感，是遠在酸、甜、苦、辣、鹹五味之上的！

因此善用「**無油無水烹調**」，鎖住營養到最高點時，除了吃到食物真實的美味，還能品嚐到食物的鮮甜，享受到天下第一幸福美味！

Natural flavor is not only eating raw or boiling without oil, salt , sugar and other sauces. The delicate flavor of its natural freshness is far beyond the five tastes ~ sweet, sour, bitter, spicy and salty.

Making a success of **"greaseless and waterless cooking"** not only locks in maximum nutrients, at the same time, it enables you to savor the natural goodness of food; and most important you will be able to taste the absolutely delicate flavor of its natural freshness which is the best in the world.

黃薑脆皮雞腿

15 分鐘 / 4人份 / 1人份178卡

icken Drumstick

15 mins/ 4 servings/ 178 cal per serving

材料：

去骨肉雞腿或玉米母雞腿2隻（約700克）

醃料：

蔥薑蒜酒汁1大匙

調味料：

黃薑粉1/2茶匙、鹽及黑胡椒各適量

作法：

1. 雞腿用檸檬皮及粗鹽逆向搓洗雞皮並沖淨，將皮下脂肪刮除，加蔥薑蒜酒汁醃約半小時，用紙巾吸乾水分，雞皮均勻抹上黃薑粉、鹽。

2. 取11吋平底煎鍋，預熱至水珠快跑，放入雞腿（雞皮貼緊鍋底，雞腿肉朝上），撒入適量的鹽及黑胡椒，加蓋。

3. 以中火煎約4分鐘至大量水氣冒出，轉中小火煎約2分鐘，開蓋並用濕布擦鍋緣，並用剪刀檢測雞腿最厚處，若快熟透時，轉中火煎約30秒至1分鐘至皮酥脆即成。

原味達人的烹調祕笈

- 傳統作法因為沒有去除多餘的皮下脂肪，熱量高又造成身體負擔。且常油炸，還必需靠許多調味料醃漬再裹粉，才能有外酥內嫩的口感。

- 用無油烹調法乾煎雞腿，享受美食又不需擔心熱量過高，因為除了無油並刮除多餘脂肪，再用檸檬皮搓洗去除雞皮殘留污垢及腥味，會有淡淡果香。將雞皮攤平貼鍋逼出油脂，再利用上下對流熱氣，保留肉質彈性，又可徹底逼出雞油及腥味，吃起來外皮香脆，多汁鮮嫩、熱量也較低。

- 可以利用乾煎兩隻雞排的空間架上大根的蘆筍（切對半），煎煮約4～5分鐘，有一鍋二菜的效果。也可建議使用控溫佳的鐵板電烤鍋，設定在400°F的恆溫乾煎約10～12分鐘。

Ingredients:

700g/2 boneless chicken leg

Marinade:

1 Tbsp mixed juice of scallion, ginger, garlic and wine

Seasonings:

1/2 tsp turmeric powder, dash salt, dash black pepper

Method:

1. Clean chicken leg by rubbing with lemon skin and sea salt, scrape off excess fat; marinate for 30 mins, pat dry, season skin evenly with salt and turmeric.

2. Preheat an 11-inch skillet till water droplets "dance quickly"; place chicken skin-side down, pressing down to ensure surface is entirely in contact with pan, sprinkle salt and black pepper, cover.

3. Pan-broil for 4 mins on medium heat till much steam appears, reduce to medium low heat for 2~3 mins; uncover, clean edge of pan with wet cloth, test for doneness by cutting the thickest part of drumstick with a pair of scissors; if almost cooked, cover and pan-broil for 1/2~1 min on medium heat till skin is crispy brown.

Culinary Tips:

· Traditional cooking methods rob food of its natural qualities and is high in calories; usually will not scrape off excess fat, the meat is floured or breaded and deep-fried to give a crispy taste; thus the need for marinating with various seasonings is necessary to restore flavor.

· Through greaseless cooking, it allows us to enjoy this dish without worrying about added calories; when drumstick begin to sear, initially sticking to the pan, it enables the excessive fat to be forced out from the skin pores and at the same time eliminating the meaty scent; however when the skin becomes crispy, the drumstick releases itself, giving a golden crispy skin that is more fragrant than floured skin, and retaining the natural juices which tenderize and flavor the chicken.

· For a more nutrition balanced dish, choice asparagus (halved) can be placed over the top as a bridge between the drumsticks for 4~5 mins; for corn-fed drumstick, electric griller is preferred and pan-broil at a constant temperature of 400°F for 10~12 mins.

香茅燉雞 25分鐘 / 4人份 / 1人份284卡

Chicken Stew with Lemon Grass

25 mins / 4 servings/ 284 cal per serving

112

材料：

母土雞腿1隻（約600克）、薑片4片、香茅2支（約70克）、洋蔥1粒（約200克）、紅辣椒1根、雞蛋馬鈴薯4顆（約280克）、紅蘿蔔約200克及香菜少許

醃料：

蔥薑蒜酒汁1大匙

調味料：

米酒2茶匙、香菇醬油50c.c.

作法：

1. 雞腿用檸檬皮及海鹽洗淨，切成4等分，加入蔥薑蒜酒汁醃約10分鐘，再吸乾水分。雞蛋馬鈴薯用軟刷洗淨不去皮；紅蘿蔔洗淨切4段（每段約50克）；香茅洗淨，切段；洋蔥去皮切4等分。

2. 取2公升燉鍋，預熱至水珠快跑，放入土雞腿（皮朝下），加蓋以中火乾爆兩面至小水氣冒出，開蓋並用濕布擦鍋緣。

3. 入薑片、香茅、洋蔥及紅辣椒，加蓋以中小火煎炒1分鐘，入米酒及香菇醬油再鋪上馬鈴薯、紅蘿蔔，加蓋，煮開至大量水氣冒出，轉小火燉煮約20分鐘至熟，搭配香菜即成。

原味達人的烹調祕笈

- 傳統作法會將所有的食材用大量熱油爆炒，再用大量的醬汁及水燉煮，容易流失營養，也降低食物的原味。
- 此道的雞腿肉塊不用油煎，直接乾爆去腥並逼出多餘油脂，再利用逼出來的油，爆香香茅及薑片，不需加水燉煮或事先汆燙馬鈴薯及胡蘿蔔，在剛好的空間運用食物產生的蒸汽，在鍋中上下對流並烹煮至熟，呈現食材原汁原味，並保留較多的營養素，更健康又美味。

Ingredients:
600g/1 wild female chicken leg (chopped into 4 pieces), 4 slices ginger, 70g/2 stalk lemon grass (segmented), 200g/1 onion (quartered), 1 red chilly, 280g/4 nugget potatoes (unpeeled), 200g carrot (50g rounds), few sprigs Chinese parsley (chopped)

Marinade:
1 Tbsp mixed juice of scallion, ginger, garlic and wine

Seasonings:
2 tsp rice wine, 50c.c. mushroom-flavored soy sauce

Method:
1. Clean chicken legs thoroughly with lemon skin and sea salt; marinate for 10 mins and pat dry. Scrub the potatoes with a brush.
2. Preheat a 2-quart pan till water droplets "dance quickly"; place in the chicken skin-side down, cover, pan-broil both sides on medium heat till little steam appears; uncover, clean edge of pan with wet cloth.
3. Add ginger, lemon grass, onion and chilly, cover, pan-fry for 1 min on medium low heat; drizzle wine and soy sauce, layer on potatoes and carrots, cover; bring to a boil till much steam appears; simmer for 20 mins on low heat till done; garnish with Chinese parsley.

Culinary Tips:
· Traditionally, extra oil is needed to fry the ingredients and much sauce and water is required to simmer this stew, losing its valuable nutrients and greatly reducing the quality of food.
· By pan-broiling the chicken, the meaty scent is eliminated, and excessive fat is forced out and used to emerge the fragrance of the spices; also without added water and parboiling the ingredients, cooking in a right size pot, covered, will allow the vapor steam to baste food in its own natural moisture and retains the natural juices and flavor; making this dish very healthy and delicious.

楓燒雞翅 10分鐘 / 3人份 / 1人份288卡

Barbecued Chicken Wings

10 mins / 3 servings/ 288 cal per serving

114

材料：

雞翅6隻、烤酥的白芝麻及檸檬皮屑少許

醃料：

蔥薑蒜酒汁2大匙

調味料：

白胡椒粉1/4茶匙、楓糖漿2大匙、楓糖醋或水果醋1大匙、檸檬汁1大匙、香菇醬油1大匙

作法：

1. 雞翅用檸檬皮及粗鹽搓洗表皮，用水沖淨並濾乾水分，加入醃料醃約30分鐘，再吸乾水分。將雞翅的關節稍微折斷並用白胡椒粉先調味，方便貼鍋攤平。

2. 取11吋平底煎鍋，預熱至水珠快跑，雞翅背面貼平鍋底壓緊，中火乾煎約3分鐘至金黃色，轉小火，開蓋，用濕布擦拭鍋緣，翻面，再以中小火煎約3分鐘。

3. 轉小火開鍋蓋，將其他調味料調勻淋在雞翅上，醬燒約30秒至1分鐘，並快速翻動雞翅至收汁成黏稠狀，雞翅上色均勻即可撒上白芝麻及檸檬皮屑即成。

原味達人的烹調祕笈

- 傳統的蜜汁烤雞翅，醃了許多醬料再燒烤，較不能逼出腥味和油脂，還會流失肉的原汁，口感不嫩。
- 雞翅乾煎法，去腥去油又封住毛細孔，只用少許的楓糖醋醬汁醬燒雞翅，口感特別鮮嫩多汁，色澤更美。
- 利用天然楓糖及楓糖醋做調味，因為楓糖漿屬唯一鹼性的糖類，糖分含量比蜂蜜低，又耐燒煮，甜度較順口。

Ingredients:
300g/6 chicken wings, dash white sesame seeds (pan-grilled), dash lemon zest

Marinade:
2 Tbsp mixed juice of scallion, ginger, garlic and wine

Seasonings:
1/4 tsp white pepper, 2 Tbsp maple syrup, 1 Tbsp maple vinegar or fruit vinegar, 1 Tbsp lemon juice, 1 Tbsp mushroom-flavored soy sauce

Method:

1. Rub chicken wings with lemon skin and sea salt, rinse and drain; marinate for 30 mins, pat dry; bend the wings at the joint so that they can lay flat on the pan to force out fat more effectively; season with white pepper.

2. Preheat an 11-inch skillet till water droplets "dance quickly"; place the back side of the wings in skillet, pressing down to ensure surface is in contact with pan, cover; pan-broil for 3 mins on medium heat till golden yellow; uncover, clean edge of pan with wet cloth, flip, cover; pan-broil for 3 mins on medium low heat.

3. Turn to low heat, uncover, clean edge of pan with wet cloth; mix the seasonings and pour over the wings, stir to deglaze the pan for sauce on medium small heat until thicken; glaze the chicken wings evenly; sprinkle sesame seeds and lemon zest before serving.

Culinary Tips:
- Traditionally, much seasoning is marinated and then roasted in the oven; thus it is unable to force out fat and eliminate its meaty scent; also the heating element in the oven will absorb its natural juices, causing it to lose its nutrients and tenderness.
- Pan-broiling the wings preserves its natural goodness which tenderizes the meat, eliminate its meaty scent and fat; only minimum sauce is needed to glaze the wings to give a vibrant color and emerge the aroma.
- Using natural maple syrup is preferred because it is the only kind of alkaline sugar; it contains less sugar than honey and the texture is smooth.

115

涼拌芝麻雞絲 20分鐘 / 4人份 / 1人份160卡

Sesame Chicken Salad
20 mins / 4 servings/ 160 cal per serving

材料：

雞里肌肉4條（約200克）、小黃瓜100克、南瓜100克、紅甜椒100克、西洋芹100克

醃料：

蔥薑蒜酒汁1大匙

調味料：

鹽少許、芝麻醬（作法見本書P.105）與檸檬優格醬（作法見本書P.101）各適量

作法：

1. 雞里肌肉用檸檬皮及海鹽洗淨，加入蔥薑蒜酒汁醃約10分鐘，再吸乾水分；將所有蔬菜切成0.2公分的細絲，用冰塊冰鎮3分鐘後，放入盤中。
2. 取8吋平底煎鍋，預熱至水珠快跑，放入雞里肌肉，撒鹽，加蓋以中火煎約2分鐘，轉小火，開蓋並用濕布擦鍋緣，翻面撒鹽，以中小火續煎約2分鐘；取出，用手剝成粗絲，鋪在作法1材料上。
3. 食用時，可將雞絲拌入適量的芝麻醬，而蔬菜絲則搭配檸檬優格醬，一次品嚐不同風味的口感。

原味達人的烹調祕笈

- 這道菜最大關鍵在於雞絲作法，傳統雞絲是先將肉放入滾水汆燙至熟，肉質易縮，還略帶雞腥味，且肉汁已流入滾水中，剝成細絲後，口感較乾澀。
- 蔬菜絲使用冰塊冰鎮，而不以傳統作法放入冰水中浸泡，可保留蔬菜原有的甜分及水分。
- 利用蔥薑蒜酒汁使肉質鮮美外，再採用乾煎法煮到剛好熟是最嫩又最多汁。用手順著肉紋剝成薄片狀，口感較不乾澀，再搭配清脆的蔬菜絲，讓味蕾體驗多層次的幸福。

Ingredients:

200g chicken breast tenderloin, 100g cucumber, 100g pumpkin, 100g red bell pepper, 100g celery

Marinade:

1 Tbsp mixed juice of scallion, ginger, garlic and wine

Seasonings:

dash salt, dash sesame sauce (see page105) or dash lemon yogurt dressing (see page101)

Method:

1. Clean chicken with lemon skin and sea salt, marinate for 10 mins, pat dry; cut all ingredients except chicken into 0.2 cm thin shreds, chill the veggies on ice for 3 mins, drain and serve on a plate.
2. Preheat an 8-inch skillet till water droplets "dance quickly"; place chicken in skillet lightly, sprinkle salt, cover; pan-broil on medium heat for 2 mins; uncover, clean edge of pan with wet cloth, flip, sprinkle salt, cover; pan-broil for 2 mins on medium low heat; tear into shreds manually; spread them on the vegetable salad in step 1.
3. Before serving, drizzle sesame sauce or lemon yogurt dressing over the chicken.

Culinary Tips:

- The main challenge of this appetizer is the preparation of the chicken breast tenderloin which have no natural oil; traditionally, chicken is parboiled which causes meat shrinkage and sometimes leaves it with a meaty scent; also, the natural juices is lost and its taste is dry and bland.
- The shredded vegetables in this salad are chilled on ice rather than soaking in iced water; this prevents the vegetables from losing their natural sweetness and moisture.
- Pan- broiling the chicken and retaining the natural juices which tenderize and flavor the meat with perfect timing; makes an excellent and delightful appetizer, when served along with the chilled crunchy salad.

117

迷迭香嫩煎梅花豬排

10分鐘 / 3人份 / 1人份359卡

Grilled Pork Picnic Steak with Rosemary　10 mins / 3 servings/ 359 cal per serving

118

材料：

梅花肉排360克、迷迭香葉少許

醃料：

蔥薑蒜酒汁1大匙

調味料：

鹽及黑胡椒適量、黑胡椒蘑菇醬（作法見本書P.107）或檸檬辣椒醬（作法見本書P.104）適量

作法：

1. 梅花肉排用檸檬皮及海鹽洗淨，切成1.5公分厚度的肉排3片，加蔥薑蒜酒汁醃約半小時後，再吸乾水分。
2. 取11吋平底煎鍋，預熱至水珠快跑，梅花肉排貼緊鍋底，撒入鹽及黑胡椒，加蓋以中火煎約2分半鐘至金黃色。
3. 轉小火，開蓋並用濕布擦鍋緣，翻面，撒入鹽、黑胡椒粉及迷迭香，加蓋以中小火續煎約2分半鐘至需要的熟度即成。食用時搭配黑胡椒蘑菇醬，或切片沾檸檬辣椒醬。

原味達人的烹調祕笈

● 傳統的豬排料理，為達鮮嫩口感，通常會泡嫩精、用鐵鎚敲打，再醃調味料裹粉油炸，易失去肉的原味，口感較油膩，熱量也高。

● 無油烹調法是以乾爆法去除腥味，逼出油脂，封住表面的毛細孔使肉汁不流失，又煎烤至剛好熟透，保持鮮嫩口感。這樣的料理方式比傳統更容易又方便，適合職業婦女。

Ingredients:

360g/3 slices pork picnic steak (1.5 cm thick), dash rosemary

Marinade:

1 Tbsp mixed juice of scallion, ginger, garlic and wine

Seasonings:

dash salt, dash black pepper, black pepper mushroom sauce (see page 107) or lemon chilly sauce (see page104)

Method:

1. Clean pork with lemon skin and sea salt, rinse and drain; marinate for 30 mins, pat dry.
2. Preheat an 11-inch skillet till water droplets "dance quickly"; place pork steak evenly in skillet lightly, pressing down to ensure surface is in contact with pan, sprinkle salt and black pepper, cover; pan-broil for 2.5 mins on medium heat till golden yellow.
3. Turn to low heat, uncover, clean edge of pan with wet cloth; flip, sprinkle salt, pepper and rosemary, cover; pan-broil for 2.5 mins on medium low heat until desired doneness; can be served with black pepper mushroom sauce or lemon chilly sauce as dip.

Culinary Tips:

· Traditional cooking methods rob food of its natural qualities and is high in calories; usually involve pounding, marinating with tenderizer and sauces, breading it and deep-fried, resulting in greasy pork.

· Using the greaseless method, food shrinkage is greatly reduced, the nutrients are locked into food, keeping the aroma in the pan; and giving a natural tenderness with perfect timing. This easy-to-prepare cooking method is much simpler compared to traditional cooking, and it is a convenient recipe for working women.

向食物 借油 原味料理
Greaseless Cooking

脫脂楓糖醋燒小排

50分鐘 / 6人份 / 1人份245卡

50 mins / 6 servings/ 245 cal per serving

120

豬肉 | Pork

材料：

五花肉豬小排600克、烤香的白芝麻2茶匙、巴西里末少許

醃料：

蔥薑蒜酒汁2茶匙、香油1/2茶匙、白胡椒粉1/4茶匙、香菇醬油1茶匙

調味料：

冰糖1茶匙、楓糖漿1大匙、楓糖醋2大匙、黑醋3/4杯、過濾水2大匙

作法：

1. 豬小排選肥肉少者佳，切成6塊並去皮及皮下脂肪，用檸檬皮及海鹽洗淨並擦乾水分，入蔥薑蒜酒汁醃約30分鐘，再加入其它醃料醃約10分鐘。
2. 取2公升燉鍋，預熱至水珠比慢跑快一點，將小排肥油較多面貼鍋乾煎，加蓋，以中火乾煎至小水氣冒出，開蓋並用濕布擦鍋緣，轉中小火繼續乾煎其餘3面呈金黃色。
3. 加入調味料以中火煮沸，轉小火續煮約30分鐘，每10分鐘翻面一次。以食物刺針檢測小排熟度，再轉中小火繼續燒煮排骨至油醋分離，收汁呈紅橙色，撒入烤香的白芝麻及巴西里末即成。

原味達人的烹調祕笈

- 傳統烹調法是先將豬小排以高溫過油，甚至裹粉油炸，如此作法不僅無法逼出油脂，還會有回鍋油及熱量增加的問題。
- 利用豬小排本身的油分乾煎，吃起來較清爽不膩又能逼出油脂；加上烹煮的溫度與時間控制得當，入口骨和肉便會分離，肉質鮮嫩卻帶咬勁，色香俱佳，一點都不輸給傳統的宴客料理。

Ingredients:
600g/6 pieces pork spareribs, 2 tsp white sesame seeds (pan-grilled), dash English parsley (minced)

Marinade:
2 tsp mixed juice of scallion, ginger, garlic and wine, 1/2 tsp sesame oil, 1/2 tsp white pepper, 1 tsp mushroom-flavored soy sauce

Seasonings:
1 tsp rock sugar, 1 Tbsp maple syrup, 2 Tbsp maple vinegar, 3/4 cup dark vinegar, 2 Tbsp filtered water

Method:

1. Select not too fatty spareribs, request butcher to cut off skin and fat beneath skin; clean spareribs with lemon skin and sea salt, rinse and drain; marinate with mixed juice for 30 mins, pat dry, then marinate with other seasonings for 10 mins.
2. Preheat a 2-quart pan till water droplets "dance medium slow-quick"; lay out the fattening part of the ribs in pan, cover; pan-broil on medium heat till little steam appears, uncover, clean edge of pan with wet cloth; continue with the same for the other 3 sides on medium low heat till golden yellow with the lid on.
3. Add seasonings, bring to a boil on medium heat; simmer for 30 mins, occasionally flipping the ribs every 10 mins; check the tenderness with cooking pin, turn to medium low heat; cook until the sauce liquidize and the grease and vinegar separates; sprinkle sesame seeds and English parsley.

121

Culinary Tips:
- Traditionally, ribs are floured and deep-fried, disabling the fat to be forced out; also starch is added to thicken the sauce.
- Cooked with perfect timing through the greaseless method, the non-greasy taste of the ribs is chewy and tender; the meaty scent is eliminated and the seasonings intensify its taste while preserving the natural goodness.

和風銀芽肉絲

12分鐘 / 4人份 / 1人份282卡

Shredded Pork and Sprouts

12 mins / 4 servings/ 282 cal per serving

材料：

霜降豬肉1片（約250克）、豆芽菜250克、青蒜絲80克、紅辣椒絲適量

醃料：

蔥薑蒜酒汁1.5茶匙

調味料：

黃砂糖1茶匙、味醂1大匙、香菇醬油1.5大匙

作法：

1. 霜降豬肉用檸檬皮及海鹽洗淨，醃蔥薑蒜酒汁約30分鐘，再吸乾水分，肉的四邊稍微切開以防捲曲；豆芽菜洗淨再去頭尾，下鍋前用水潤濕，放入11吋平底煎鍋，以中火煮約1分鐘，稍微搖動鍋子至小水氣冒出，拌炒起鍋裝盤。

2. 取8吋平底煎鍋，預熱至水珠快跑，放入霜降豬肉片，加蓋以中火煎約3分鐘，轉小火，開蓋，翻面以中小火繼續煎3分鐘至熟透，取出，斜切成0.5公分的片狀，再切成0.5公分細絲。

3. 倒出鍋中逼出的油，再放入調味料，轉中小火煮沸，加入肉絲拌炒至收汁，熄火，續入青蒜絲及紅辣椒絲拌勻，鋪於豆芽菜上面即成。

原味達人的烹調祕笈

- 傳統烹調霜降豬肉絲，會用較多的調味料醃入味，再入油鍋中過油，表面會有一層油，很難品嚐到霜降豬肉絲的原味。

- 霜降豬肉表面有豐富油脂，呈薄片狀，先利用乾爆法逼出油脂並封住表面毛細孔，再切成絲狀並調味，不需用油及太多的醬料就可保留其特殊的肉汁及甜分，口感香Q帶勁。

- 豆芽菜經過清洗後，表面會殘留水珠；利用無水烹調的原理，使熱氣上下對流均勻，只要冷鍋開火煮的剛好熟時，口感更清脆。此道佳餚可直接食用，也可搭配全麥潤餅皮、青蒜絲及廣東A菜，做成春捲食用。

Ingredients:

250g pork neck bacon, 250g bean sprouts (rip off ends), 80g leeks (shredded), 1 red chilly (shredded)

Marinade:

1.5 tsp mixed juice of scallion, ginger, garlic and wine

Seasonings:

1 tsp cane sugar, 1 Tbsp mirin, 1.5 Tbsp mushroom-flavored soy sauce

Method:

1. Clean pork with lemon skin and sea salt, rinse and drain; marinate for 30 mins, pat dry; give a small slit on 4 sides of pork to prevent curling; rinse bean sprouts before cooking, place them in an 11-inch skillet; cook for 1 min on medium heat shaking the skillet slightly till little steam appears, stir till cook evenly and remove.

2. Preheat an 8-inch skillet till water droplets "dance quickly"; place pork in, cover; pan-broil for 3 mins on medium heat; uncover, clean edge of pan with wet cloth, flip; pan-broil for 3 mins on medium low heat till done; remove, cut into 0.5 cm shreds.

3. Discard excess grease in pan, add seasonings, deglaze the pan for sauce on medium low heat; add shredded pork, give a quick stir, then stir in leeks and chilly; spread onto bean sprouts and serve.

123

Culinary Tips:

· Traditionally, pork is shredded, seasoned and quick-fried in hot oil; resulting in greasy pork.
· With greaseless cooking, pan-broiling the pork as a whole preserves the natural moisture and sweetness; only minimum sauce is needed to fry the shredded pork with no extra oil.
· Using the waterless cooking method, the bean sprouts is basted in its own natural moisture, without parboiling, giving a crunchy refreshing taste; this dish can also be served as a wrap.

香椿醬烤羊小排

20分鐘 / 4人份 / 1人份288卡

dar Pesto

20 mins / 4 servings/ 288 cal per serving

材料：

美國羊小排2大塊（約600克）

醃料：

蔥薑蒜酒汁1大匙

調味料：

黑胡椒及鹽適量、香椿青醬1/2杯（作法見本書
P.103）

作法：

1. 羊小排用粗鹽及檸檬皮洗淨，再吸乾水分，加
   入蔥薑蒜酒汁醃約30分鐘。

2. 取11吋平底煎鍋，預熱至水珠快跑狀（或鐵板
   電烤鍋預熱至400°F），放入均勻抹鹽的羊小
   排，多油面朝下貼緊鍋底，加蓋以中火煎約2分
   鐘至金黃色，開蓋並用濕布擦鍋緣，轉小火，
   翻面，撒黑胡椒，加蓋以中小火續煎兩面各3分
   鐘（或鐵板電烤鍋轉325°F），轉小火。

3. 將羊小排再切成8小片，加蓋，剖面貼鍋以中
   火煎約2分鐘（或鐵板電烤鍋預熱至400°F）
   至金黃色，轉小火，翻面塗抹少許的香椿醬
   於羊排上，加蓋轉中小火（或鐵板電烤鍋轉
   325°F），煎約2分鐘至需要的熟度。（若不想
   過熟，就不要切成小片，以中小火繼續煎3分
   鐘。）

原味達人的烹調祕笈

- 傳統羊小排若用烤箱烹調完成，仍會有羊腥味，
  肉質會縮掉且口感較乾澀，故習慣在醬料中加嫩
  精及奶油。

- 羊小排用蔥薑蒜酒汁醃，可增添鮮美的甜度，再
  用整塊貼鍋乾爆逼出多餘的油脂，可完整去除羊
  腥味又封住表面毛細孔，留住鮮肉汁，呈現肉質
  鮮嫩原味的口感。若用自製的香椿醬提味，料理
  將有加分的效果。

Ingredients:

600g/2 American Lamb Rack

Marinade:

1 Tbsp mixed juice of scallion, ginger, garlic and
wine

Seasonings:

dash black pepper, dash salt, 1/2 cup cedar pesto
(see page 103)

Method:

1. Clean lamb rack with lemon skin and sea salt,
   rinse and drain; marinate for 30 mins, pat dry.

2. Preheat an 11-inch skillet till water droplets
   "dance quickly" (for electric griller, preheat to
   400°F); season with salt, place the fattening
   part of lamb rack in pan, cover; pan-broil for 3
   mins on medium heat (griller at 400°F) till golden
   yellow; uncover, clean edge of pan with wet
   cloth, flip, sprinkle pepper, cover; pan-broil for 2
   mins on medium low heat (griller at 325°F), turn
   to low heat.

3. Slice and lay out riblets, cover; pan-broil for 2
   mins on medium heat (griller at 400°F) till golden
   yellow; turn to low heat, flip, spread lamb with
   pesto on top, cover; pan-broil the other side for
   2 mins on medium low heat (griller at 325°F)
   till desired doneness; (for medium doneness, do
   not slice into riblets, pan-broil the fattening side
   of lamb rack for another 3 mins on medium low
   heat).

Culinary Tips:

· Roasting lamb rack in an oven tend to cause
  meat shrinkage and is unable to eliminate the
  meaty scent; also the lamb taste drier and
  tougher, so the need for tenderizer and added
  butter is necessary to restore flavor.

· With greaseless cooking, the meaty scent and
  excessive fat is eliminated; the natural juices
  are retained for a tender taste; the cedar pesto
  complements well with grilled lamb, intensifying its
  flavor.

原味牛肩肉骨茶湯

60分鐘 / 8人份 / 1人份152卡

60 mins / 8 servings/ 152 cal per serving

126

材料：

台灣黃牛肩肉1條（約800克）、不含藥味肉骨茶包1包、帶膜蒜頭2大球、大蘑菇8粒（約800克）、綠竹筍2支（約800克）、香菜段30克

作法：

1. 先剪除牛肩肉多餘油脂，用檸檬皮及海鹽洗淨並濾乾；大蘑菇洗淨；綠竹筍剝除外殼，洗淨，直切成4等分。

2. 取4公升湯鍋，預熱至水珠快跑，將牛肉較肥部位貼緊鍋底，加蓋以中火乾爆兩面至小水氣冒出，並呈金黃色，倒入熱開水1000c.c.、肉骨茶包、蒜頭，加蓋以中火煮沸，轉小火燉煮約25分鐘。

3. 取出肉骨茶包，加入熱開水1500c.c.及綠竹筍，以中火煮沸再轉小火繼續燉煮約10分鐘，放入大蘑菇，再煮沸，轉小火燉煮約8分鐘至熟透，可依個人喜好加熱水調味。取出已燉熟的牛肩肉，切成2公分厚片，將湯料平分放入碗裡，淋上肉骨茶湯汁，搭配香菜段即成。

原味達人的烹調祕笈

- 傳統烹煮牛肉習慣切小塊再以大量的湯水燉煮，造成牛肉容易縮水外，本身的甜分和營養也會被破壞。

- 每頭牛只有兩條牛肩肉，將整條的黃牛肩肉乾爆去腥去油，再加入少量滾水大塊燉煮，可讓肉質香甜；煮到剛好的熟度，再取出切厚片，較能保留肉質的嫩度及Q感。

- 大蒜整球帶膜煮，不會因為長時間的燉煮而造成湯汁渾濁。

Ingredients:

800g Taiwanese chuck flat iron beef, 1 sachet Bah Kut Teh spices (original flavor), 2 garlic bulb (peeled), 800g/8 large button mushrooms, 800g/2 bamboo shoots, 30g Chinese parsley (chopped coarsely)

Method:

1. Trim off excessive fat from beef, clean with lemon skin and sea salt, rinse and drain, pat dry; peeled bamboo shoots and cut vertically into 4 pieces.

2. Preheat a 4-quart pot till water droplets "dance quickly"; place the fattening side of beef in, cover; pan-broil both sides on medium heat till little steam appears and golden yellow; pour in 1000 c.c. hot water, add spice sachet and garlic, bring to a boil on medium heat; simmer for 25 mins on low heat.

3. Remove spice sachet, add 1500 c.c. hot water and bamboo shoots, bring to a boil, simmer for 10 mins on low heat; add mushrooms, bring to a boil and simmer for another 8 mins till done; add hot water to desired taste; remove beef, cut into slices 2 cm thick; serve hot with Chinese parsley.

Culinary Tips:

· Traditionally, beef is cut beforehand and cook in too much water, causing meat shrinkage and destroying the flavor.

· Pan-broiling the beef as a whole, not only eliminate its meaty scent and excessive fat; when simmer with minimum water with perfect timing, it retains the natural sweetness, aroma and tenderness of the beef.

· When cooking soups, peeling and separating the garlic bulb in cloves, will cause the garlic to be easily overcooked, clouding the soup.

椒鹽烤無骨牛小排

15分鐘 / 6人份 / 1人份470卡

15 mins / 6 servings/
470 cal per serving

128

材料：

加拿大無骨牛小排2片（共約800克，寬8公分，長18公分）

調味料：

玫瑰鹽或海鹽適量、黑胡椒適量

作法：

1. 牛小排用檸檬皮及海鹽洗淨，再吸乾水分。
2. 取12吋平底煎鍋，預熱至水珠快跑，將牛小排油脂處貼緊鍋底，撒適量鹽，加蓋以中火煎第一面約3分鐘（或鐵板電烤鍋預熱至400℉，煎兩面各約3.5分鐘），開蓋並用濕布擦鍋緣，翻面，撒鹽及黑胡椒粉，以中小火煎約3.5分鐘。
3. 牛小排先切成兩長條，寬約4公分，長約18公分，撒鹽及黑胡椒粉，以中小火再煎另外兩面各約3.5分鐘（或鐵板電烤鍋轉至325℉）至需要的熟度即成。

原味達人的烹調祕笈

◎ 傳統烤牛小排為達鮮嫩口感，習慣泡嫩精，用奶油煎後，再移入烤箱烤到九分熟時，外表焦黃，內部會乾澀，也比較不能瞬間把肉腥味逼掉。

◎ 使用可控溫的鐵板電烤鍋無油乾煎牛小排的效果最好，可以持續高溫煎烤不會讓牛排燒焦或乾硬。利用400℉的特定恆溫乾爆，將牛小排表面的毛細孔封住，去腥又去脂，還可鎖住香甜肉汁不外流，即便煎至九分透，口感仍然香嫩帶勁，肉質不乾澀。

Ingredients:

800g boneless Canadian beef short ribs (8 cm wide X 18 cm long)

Seasonings:

dash rose or sea salt, dash coarse black pepper

Method:

1. Clean beef with lemon skin and sea salt, rinse and drain, pat dry.
2. Preheat an 11-inch skillet till water droplets "dance quickly", place the fattening side of beef in pan, sprinkle salt, cover; pan-broil for 3 mins on medium heat (for electric griller, preheat to 400℉, pan-broil both sides for 3.5 mins); uncover, clean edge of pan with wet cloth, flip, sprinkle with salt and pepper, cover; pan-broil for 3.5 mins on medium low heat.
3. Cut beef into 2 slices, 4 cm x 18 cm; sprinkle salt and pepper, cover; pan-broil the other 2 sides for 3.5 mins on medium low heat (griller at 325℉) till desired doneness.

Culinary Tips:

· Traditionally, the beef is marinated with tenderizer, pan-fried in butter, and then grilled in an oven, losing much of its valuable nutrients.

· For best results, pan-broiling beef initially at a constant high temperature at 400℉ seals the moisture within the meat, which tenderize and flavor it; as well as eliminating its meaty scent and excessive fat when the beef begin to sear immediately and initially sticking to the pan; also the high constant temperature will emerge the fragrance but will not dry or burn the meat.

香根牛肩胛肉

10分鐘 / 2人份 / 1人份198卡

Chinese Parsley

10 mins / 2 servings/ 198 cal per serving

材料：

加拿大牛肩胛翼板肉片200克、蒜片2粒、青蒜1支（約80克）、香菜300克、紅辣椒絲1根、芥花籽油1茶匙

醃料：

香油1/2茶匙、蠔油1茶匙、香菇醬油1大匙、黑胡椒粉1/2茶匙、蔥薑蒜酒汁1茶匙

作法：

1. 牛肉洗淨，再吸乾水分，切成0.5公分肉絲，加入醃料醃約10分鐘；青蒜洗淨，對剖，斜切成細絲；香菜洗淨，取根莖部位，切成長段狀。

2. 取8吋平底煎鍋，預熱至水珠慢跑，放入油、蒜片，加蓋以小火爆香兩面各半分鐘至金黃色，取出，入牛肉絲貼鍋攤平，鋪上爆香好的蒜片，加蓋，轉中火煎約1分鐘，轉小火，開蓋並用濕布擦鍋緣，翻面，再加蓋以中小火煎約半分鐘。

3. 轉小火，開蓋入青蒜絲、香菜及紅辣椒絲，以中火快速拌炒熄火，利用餘溫快速拌炒至熟，起鍋後建議馬上食用。

原味達人的烹調祕笈

- 傳統爆炒牛肉絲習慣放入油鍋中過油或泡油，或醃太白粉再放入鍋中不停翻炒，使肉絲滑口，卻不斷吸取鍋面的油，口感較油膩。

- 無油烹調牛肉絲，是以少油煎炒，透過加蓋烹調所產生的熱氣上下對流烹煮，再翻面快炒至熟，可去腥，保留肉汁和肉香，火候時間掌控得宜時，不需嫩精，不需勾芡，口感一樣滑嫩好吃。

Ingredients:
200g Canadian beef chuck flap (0.5 cm shreds), 2 cloves garlic (sliced), 80g/1 stalk leek, 300g Chinese parsley (stemmed), 1 red chilly, 1tsp canola oil

Marinade:
1/2 tsp sesame oil, 1 tsp oyster sauce, 1 Tbsp mushroom-flavored soy sauce, 1/2 tsp black coarse pepper, 1 tsp mixed juice of scallion, ginger, garlic and wine

Method:
1. Marinate shredded beef for 10 mins; cut leek lengthwise in half, then at an angle into thin shreds.
2. Preheat an 8-inch skillet till water droplets "dance slowly"; add oil, garlic, cover; pan-fry both sides for 1/2 min each on low heat till golden yellow, remove; spread beef evenly in pan, add garlic on top, cover; pan-fry for 1 min on medium heat; turn to low heat, uncover, clean edge of pan with wet cloth, flip, cover; pan-fry for 1/2 min on medium low heat.
3. Turn to low heat, add leek, parsley stems and red chilly, give a quick stir on medium fire till doneness; serve immediately.

131

Culinary Tips:
· Traditionally, shredded beef are parboiled in hot oil and strained or quick fried with much oil, resulting in greasy beef; it is usually marinated with starch to give a smooth taste.
· With greaseless cooking, frying lean shredded beef with a little oil, covered, cooks food in its own natural moisture and keeping the aroma in the pan; with right heat and perfect timing, the natural juices and flavor will be retained with a delicate and tender taste.

椒麻黃魚

10分鐘 / 4人份 / 1人份132卡

10mins / 4 servings / 132 cal per servin

材料：

黃魚2尾（約700克，10吋長）

醃料：

蔥薑蒜酒汁2大匙

調味料：

花椒鹽1茶匙

花椒鹽DIY：

材料有礦物鹽100克、花椒粒50克。取1公升鍋，預熱至水珠慢跑，加蓋以小火炒花椒粒5分鐘後取出磨碎；入礦物鹽於鍋中，加蓋以小火慢慢炒約3分鐘，再放入磨碎的花椒一起炒約3分鐘即熄火，冷卻後裝罐保存。

作法：

1. 取出黃魚內臟，再用檸檬皮及粗鹽搓洗後沖淨，濾乾水分，加入蔥薑蒜酒汁醃約30分鐘，再擦乾。

2. 取11吋平底煎鍋，預熱至水珠快跑，將黃魚貼平鍋底並輕壓（魚肚面向鍋邊，頭尾交錯放）至油脂逼出，加蓋。

3. 以中火乾煎約3分鐘至小水氣冒出呈金黃色，開蓋並用濕布擦鍋緣，翻面，均勻撒上花椒鹽，加蓋以中小火續煎約3分鐘至熟，起鍋盛盤沾食花椒鹽。

原味達人的烹調祕笈

- 傳統為保持魚皮酥脆金黃，會以多油煎炸；為求快熟，還會在魚腹部位斜劃二三刀，不但流失肉汁又噴油，易吸取較多油脂；怕魚皮黏鍋，則用不沾鍋，卻無法去腥去脂，對健康不好又無法吃到魚肉的鮮甜。

- 無油烹調黃魚須選擇魚腹及魚尾厚度均等，魚尾就不會過熟；乾煎時，貼緊鍋子逼出油脂及腥味，達到魚皮酥香，魚肉滑嫩多汁。

- 可依個人喜好搭配茄汁蝦皮辣醬、蒜片金桔醬。（作法可見本書P.102及P.106）

Ingredients:

700g/2 yellow fish (10 inch long)

Marinade:

2 Tbsp mixed juice of scallion, ginger, garlic and wine,

Seasonings:

1 tsp Szechwan peppercorn mineral salt

Preparation of Szechwan peppercorn mineral salt:

Ingredients: 100g mineral salt, 50g Szechwan peppercorn; preheat a 1-quart pan till water droplets "dance slowly"; pan-grille peppercorn with lid on for 5 mins on low heat, remove and grind; fry salt for 3 mins on low heat; then add in grinded peppercorn and stir for another 3 mins..

Method:

1. Discard the innards of the fish, rub fish with lemon skin and sea salt, rinse and drain; marinate for 30 mins, pat dry.

2. Preheat an 11- inch skillet till water droplets "dance quickly"; place fish flat and slightly curved with fish belly near edge of pan and in an alternate order of head to tail; lightly press down till traces of fat is forced out; the fish will initially stick to the pan to eliminate the fishy scent and excess fat, cover.

3. Pan-broil for 3 mins on medium heat until little steam appears and the fish is golden yellow; uncover, clean edge of pan with wet cloth, flip, sprinkle Szechwan peppercorn salt, cover; pan-broil for 3 mins on medium low heat till done; serve with dash peppercorn rose salt as dip.

Culinary Tips:

- Traditionally, cuts have to be made at belly, fish is floured and deep-fried in hot oil to give a crispy taste; or fried in a non-stick pan to reduce the use of oil; however, the fish leaves the surface of the pan during the frying process disabling the fishy scent and fat to be forced out; both methods will absorb extra oil and lose its valuable nutrients and natural juices.

- With greaseless cooking, the fishes should be consistent in size; the thickness of belly and tail should not be of much difference so as to preserve the nutrients of whole fish; by pan-broiling the fish without extra oil, it will still create a crispy, golden yellow appearance with a tender, juicy and smooth texture that is more fragrant than traditional methods.

- The spicy tomato dried small shrimp sauce or garlic kumquat sauce sauce can be used as suitable dips instead.(see page 102 and page 106)

香草魚排佐蘋果番茄沙沙醬
10分鐘 / 4人份 / 1人份105卡

Codfish with Apple Tomato Salsa
10 mins / 4 servings / 105 cal per serving

材料：

圓鱈1片（500克，厚約2.5公分）、帶皮蘋果150克、檸檬汁1茶匙、番茄沙沙醬1/2杯（作法見本書P.100）

醃料：

蔥薑蒜酒汁2茶匙

調味料：

礦物鹽及義大利綜合香料適量

作法：

1. 圓鱈魚刮除表皮的魚鱗並用檸檬皮及海鹽洗淨，加入蔥薑蒜酒汁醃約30分鐘後，再吸乾表面的水分；蘋果洗淨，帶皮切成小丁，加入檸檬汁及番茄沙沙醬一起拌勻。

2. 取8吋平底煎鍋預熱至水珠快跑，放入圓鱈魚貼緊鍋底至油脂逼出，撒入少許礦物鹽，加蓋，以中火煎約3分鐘至小水氣冒出。

3. 轉小火並用濕布擦鍋緣，翻面，撒入少許礦物鹽及綜合香料，加蓋以中小火續煎約3分鐘至需要的熟度，起鍋。將作法1的蘋果沙沙醬以模型扣成型再倒入盤中，上面放入煎好的魚排即成。

原味達人的烹調祕笈

- 傳統中式料理鱈魚方式，習慣切薄片，又因肉質太嫩容易沾鍋散開，所以大部分會裹粉再放入油鍋中油炸，口感較油膩；而西式料理厚片鱈魚，則是用奶油先煎至表面呈金黃色，再移入烤箱烤熟，表面肉質較乾，魚肉原汁及營養素容易流失。

- 採用無油乾爆法，可徹底去腥，鎖住原汁；掌握火候溫度得宜，將魚肉煎到剛好的熟度時，肉質滑Q，口感像頂級鮮干貝。

- 蘋果切開容易氧化變色，大多會放入鹽水中浸泡，而流失甜分；改用檸檬汁拌勻，還多了檸檬的清香味，吃起來微酸微甜，口感更佳。

Ingredients:

500g/1 slice codfish fillet (2.5 cm thick), 150g apple (unpeeled, diced), 1 tsp lemon juice, 1/2 cup tomato salsa sauce (see page 100)

Marinade:

2 tsp of mixed juice of scallion, ginger, garlic and wine

Seasonings:

dash mineral salt, dash mixed Italian spices

Method:

1. Scrape off scales from fish skin, clean fish with lemon skin and sea salt, rinse and drain; marinate for 30 mins, pat dry; mixed well diced apple with lemon juice and then tomato salsa sauce.

2. Preheat an 8-inch skillet till water droplets "dance quickly"; place fillet in pan, lightly pressing down until traces of fat is forced out, sprinkle salt, cover; pan-broil for 3 mins on medium heat till little steam appears.

3. Turn to low heat, clean edge of pan with wet cloth, flip, sprinkle salt and Italian spices, cover; pan-broil for 3 mins on medium low heat until desired doneness; remove; shape mixed salsa in step 1 in a ring mould on a plate; quartered the fillet and serve on top of salsa.

Culinary Tips:

- Traditionally, codfish is sliced thin, floured to prevent it from falling apart, and deep-fried adding excessive calories; or pan-fried with butter and then baked in oven, robbing the fish of its valuable nutrients and natural juices.

- Pan-broiling without extra oil is the best cooking method for this high quality fish, as it seals the natural juices within, eliminates the fishy scent and excessive fat; giving it a succulent and delicate aroma, and resembling the texture of fresh scallop.

- When apple is diced, it usually changes color due to oxidation; thus it is soaked in salt water, losing its natural sweetness; lemon which is a healthier choice, helps to enhance its flavor with a light lemon aroma.

135

清烤魩仔魚vs.醬燒丁香魚

15分鐘 / 6人份 / 1人份168 vs.137卡

al Fish vs. Sauté of Silver Anchovy with Peanuts

136

清烤魩仔魚

材料：

魩仔魚200克、蒜頭2顆

作法：

1. 先將魩仔魚放在濾網中用過濾水洗淨，再吸乾水分；蒜頭輕拍備用。
2. 取8吋平底煎鍋，預熱至水珠比慢跑快一點，放入魩仔魚鋪平，蒜頭放在魚上，加蓋以中火烤兩面各約2分鐘至金黃色。
3. 轉小火，開蓋翻面，加蓋以中小火續烤約2分鐘至熟。

醬燒丁香魚

材料：

丁香魚乾150克、蒜片20克、紅辣椒片15克、青蔥末60克、烤香花生粒60克、芥花油1茶匙

調味料：

香菇醬油1.5茶匙、細砂糖1/2茶匙

作法：

1. 先將丁香魚乾洗淨，再吸乾水分。
2. 取8吋平底煎鍋，預熱至水珠慢跑，入少許油以小火爆香蒜片兩面各30秒至金黃色，取出。放入丁香魚，爆香的蒜片放在魚上，加蓋，以中火煎2分鐘至金黃色。
3. 轉小火，開蓋翻面，加蓋以中小火煎約2分鐘，入調味料、紅辣椒片及青蔥末，以小火快速拌炒約1分鐘即成，搭配花生粒食用。

原味達人的烹調祕笈

- 傳統料理魩仔魚會用多油煎炒，丁香魚乾則用油炸，為增加酥脆口感，甚至會多炸二次，利用熱油的溫度，促使水分消失，品嚐不到原味，熱量也較高。
- 烤花生以控溫佳的鐵板電烤鍋先以400℉烤5分鐘，再轉325℉續烤約40～50分鐘至剛好熟透時，花生香脆清甜，較不燥熱。

Pan-Fry Larval Fish

Ingredients:

200g larval fish, 2 cloves garlic (crushed)

Method:

1. Place larval fish in a strainer, rinse and drain, pat dry.
2. Preheat an 8-inch skillet till water droplets "dance medium slow-quick"; spread fish evenly on pan, add garlic on top, cover; pan-broil for 2 mins on medium heat till golden yellow.
3. Turn to low heat, uncover, flip, cover; pan-broil for 2 mins on medium low heat till done.

Saute of Silver Anchovy with Peanuts

Ingredients:

150g Silver Anchovy, 20g/2 cloves garlic (sliced), 15g red chilly (sliced), 60g scallion (minced), 60g peanuts (pan-grilled), 1 tsp canola oil

Seasonings:

1.5 tsp mushroom-flavored soy sauce, 1/2 tsp cane sugar

Method:

1. Place silver anchovy in a strainer, rinse and drain, pat dry.
2. Preheat an 8-inch skillet till water droplets "dance slowly"; add oil, garlic, pan-fry both sides for 1/2 min each on low heat till golden yellow, remove; add silver anchovy, garlic on top, cover; pan-fry for 2 mins on medium heat till golden yellow.
3. Turn to low heat, uncover, flip, cover; pan-fry for 2 mins on medium low heat; add seasonings, chilly and scallion, give a quick stir to mix well on medium heat; serve with peanuts.

137

Culinary Tips:

- Traditionally, larval fish is fried in oil and anchovy is deep-fried; to enhance the crispiness, sometimes it is fried more than once, so that the moisture is released; however, it is high in calories and loses its original flavor and nutrition.
- Raw peanuts (rinse before frying) can be pan-grilled on electric griller at 400℉ for 5 mins with the lid, then at 325℉, occasionally shaking the pan, for 40~50 mins till done; it is nutritious, crunchy and naturally flavor-filled.

洋蔥煎蛋

10分鐘 / 4人份 / 1人份115卡

10 mins / 4 servings / 115 cal per serving

材料：

雞蛋3顆（約200克）、洋蔥120克、紅蘿蔔60克、青蔥末30克、芥花油1茶匙

調味料：

鹽1/4茶匙、白胡椒粉1/8茶匙

作法：

1. 雞蛋打散，加入調味料拌勻。洋蔥橫切成0.3公分細絲，胡蘿蔔切成0.2公分細絲。

2. 取11吋平底煎鍋，預熱至水珠比慢跑快一點，入油抹勻鍋底，依序鋪入洋蔥絲、胡蘿蔔絲、青蔥末，再均勻地覆蓋蛋液，加蓋，以中小火煎約1分鐘，轉中火續煎約1分鐘至底部呈金黃色（約八分熟）。

3. 轉小火，開蓋並用濕布擦鍋緣，用剪刀剪成4等分，翻面，不要用鍋鏟壓出蛋液，加蓋以中小火續煎1分鐘全熟即成。

原味達人的烹調祕笈

- 傳統作法因洋蔥甜分較多，容易燒焦，就用多油拌炒去除洋蔥辛辣味；煎蛋時，為了口感滑嫩，常用過多的油，又常用鏟子壓出未熟的蛋液，反而容易吸收更多的油脂更油膩。

- 原味烹調強調依食物的量，挑選適中的鍋具，預熱及火候得當，再依食材屬性依序放入鍋中，就可以用最少的油爆香及去除食材的腥味，又可利用食物烹調時產生的蒸汽同時煮得剛好熟透，吃到原汁原味。

- 洋蔥切法會影響口感，順著紋路切，吃起來較香脆，不適用此道料理；而逆著紋路切較快熟，口感軟嫩香甜。

Ingredients:

200g/3 eggs, 120g onion, 60g carrot (0.2 cm shreds), 30g scallion (minced), 1tsp canola oil

Seasonings:

1/4 tsp salt, 1/8 tsp white pepper

Method:

1. Beat the eggs, add seasonings, whisk eggs manually; cut onion vertically into 0.3 cm shreds.

2. Preheat an 11-inch skillet till water droplets "dance medium slow-quick"; grease pan evenly, spread onion evenly on pan, lay out carrot and scallion on top, and then drizzle egg mixture over the top, cover; pan-fry for 1min on medium low heat, then for 1 min on medium heat until medium-well done and golden yellow.

3. Turn to low heat, uncover, clean edge of pan with wet cloth, quartered with a pair of scissors, flip, do not add pressure to the egg with a slice, cover; pan-fry for 1 min on medium low heat till done.

Culinary Tips:

- Traditionally, onions which have a high sugar content, is quick fried with much oil to prevent burning and to eliminate its spicy scent; for omelet, usually much oil is used to retain its tenderness and during the process, the egg mixture is pressed down to force out uncooked eggs, resulting in greasy omelet.

- The natural low fat cooking emphasizes selecting the right size pan, pan heat, correct control of heat during cooking process and placing the ingredients in the right sequence; also the use of oil is reduced to the minimum by adding only once and it still retains the fragrant, tender and naturally flavor-filled taste with no spicy scent of the onions.

- How the onion is cut will affect the texture and taste; if cut horizontally against the lines, the onion juices will be partly released after cutting and taste softer; if cut vertically along the lines, the onion will taste crispier and will not be that juicy, which is more suitable for this dish.

鐵板蔬菜豆腐

15分鐘 / 4人份 / 1人份203卡

Vegetarian Tofu Teppanyaki

15 mins / 4 servings / 203 cal per serving

140

材料：

雞蛋豆腐2盒（約600克）、洋蔥150克、新鮮香菇150克、番茄300克、西洋芹1根（約100克）、荷蘭豆夾80克、紅黃甜椒各80克、芥花籽油1茶匙

調味料：

醬油1/4杯、酒1茶匙、高湯1/4杯

作法：

1. 食材分別洗淨；雞蛋豆腐濾乾，每盒切成3大塊（約3公分厚）；洋蔥、新鮮香菇、番茄、甜椒切塊（約2公分寬）；西洋芹去除老筋，切2公分斜段。

2. 取11吋平底煎鍋，預熱至水珠比慢跑快一點，入油抹勻鍋底，加入雞蛋豆腐，加蓋以中火煎約3分鐘，轉小火，開蓋，並用濕布擦鍋緣，翻面，加蓋以中小火煎約1分鐘。

3. 續入洋蔥於豆腐空隙，再將香菇、番茄擺放在豆腐上面，加蓋，以中小火煎約1分30秒，加入西洋芹、荷蘭豆莢及甜椒再淋調味料於豆腐上，加蓋以中火煮開，轉小火煮約1分鐘至熟。

原味達人的烹調祕笈

- 傳統料理紅燒豆腐，大多把豆腐放入油鍋炸至酥脆，使豆腐的水分流失熱油中，容易產生油爆，又有致癌的疑慮；蔬菜另外過油又用過多調味料燜煮入味再勾芡，豆腐口感不夠滑嫩，蔬菜容易過熟油膩。

- 豆腐本身無油，故需以恆溫少油，並加蓋清煎大塊豆腐，不但煎成金黃酥香，也可保留豆腐原味，呈現類似豆花口感。利用食材本身烹煮時產生的蒸汽，將番茄、菇類、芹菜等一起燜透，再用少量醬汁瞬間燴到豆腐上，縮短紅燒時間，並保留每樣食材的營養素。

Ingredients:

600g/2 boxes egg tofu, 150g onion, 150g fresh black mushrooms, 300g tomato, 100g celery (devein), 80g peapods (devein), 80g each red and yellow bell peppers, 1 tsp canola oil

Seasonings:

1/4 cup mushroom-flavored soy sauce, 1 tsp wine, 1/4 cup chicken broth

Method:

1. Rinse and drain, sliced tofu 3 cm thick; cut onion, mushrooms, tomato, celery and bell peppers into 2 cm wide slices; pry up onion slices with 3 layers intact.

2. Preheat an 11-inch skillet till water droplets "dance medium slow-quick"; grease pan evenly with oil, lay out tofu evenly, cover; pan-fry for 3 mins on medium heat; turn to low heat, uncover, clean edge of pan with wet cloth, flip, cover; pan-fry for 1 min on medium low heat.

3. Spread onion on pan between tofu, top with mushrooms and tomato, cover; pan-fry for 1.5 mins on medium low heat; add celery, peapods, bell peppers, drizzle seasonings over the top, cover; bring to a boil and simmer for 1min on low heat till done.

Culinary Tips:

- Traditionally, tofu is sliced thin and deep-fried until golden brown and oil splatters easily, creating a mess and risk of lung cancer; the veggies are then stir-fried separately with much oil and simmered and immersed with much seasonings and starch; thus the veggies tend to absorb too much oil and sauce, and become overcooked, losing its valuable nutrients.

- As tofu has no natural oil, in order to enhance its flavor, the tofu is sliced thick, fried with a little oil, covered, to fragrant the tofu on the surface and within, leaving it with a tender pudding-like texture; the veggies cook completely in its own natural moisture without added oil, locking the natural goodness into every ingredient and keeping the aroma in the pan.

向食物借油　原味料理
Greaseless Cooking

椒鹽烤明蝦 10分鐘 / 3人份 / 1人份130卡

Pan Grilled Peppery Salt King Prawns

10 mins / 3 servings / 130 cal per serving

142

蝦蟹貝螺類 | Shellfish

材料：

明蝦3隻（約360克）、小番茄16顆、菊苣1小顆

醃料：

蔥薑蒜酒汁1大匙

調味料：

粗鹽3大匙及黑胡椒1茶匙

作法：

1. 明蝦用牙籤去除腸泥，剪除蝦腳、蝦頭並洗淨，加入蔥薑蒜酒汁醃約30分鐘，再擦乾水分，黑胡椒均勻撒在蝦腹。

2. 取11吋平底煎鍋，預熱至水珠比慢跑快一點，均勻撒入粗鹽於鍋底，擺入明蝦（蝦腹朝下，頭尾交錯站立），加蓋。

3. 以中火烤約3分鐘至小水氣冒出，轉中小火繼續烤約3～5分鐘至熟（或用鐵板電烤鍋調至400°F，烤6～8分鐘至熟），以食物刺針測試熟度，搭配小番茄、菊苣即成。

原味達人的烹調祕笈

- 傳統烤明蝦為了讓蝦肉口感脆Q，常泡小蘇打或硼砂，再放在網架上烤；或在表面抹奶油入烤箱烤熟，但口感常太乾或太油，加上肉質過熟導致蝦身彎曲或縮水，流失養分又無法享受甜美的原味。

- 原味烹調先將蝦腳剪掉，一來可貼在鍋面站穩，二來可去除海腥味；黑胡椒撒在蝦腹處，味道容易進到蝦肉，還能在最短時間讓肉質快熟而不致變形。

- 高級食材若要吃原味，無油乾烤是最好的方法，可保留風味、營養素、消化酶及菜色，品嚐到蝦肉的鮮美及肉質彈Q的口感。此道佳餚可直接食用，或依個人喜好搭配檸檬辣椒醬，做不同口味的變化。

Ingredients:

360g/3 king prawns, 16 small tomatoes, 1 branch leafy greens

Marinade:

1 Tbsp mixed juice of scallion, ginger, garlic and wine

Seasonings:

3 Tbsp coarse sea salt, 1 tsp coarse black pepper

Method:

1. Remove intestinal vein of prawns with toothpick, trim off legs and sharp ends of head, wash clean; marinate for 30 mins, pat dry; sprinkle black pepper evenly on stomach.

2. Preheat an 11-inch skillet till water droplets "dance medium slow-quick"; sprinkle sea salt evenly on pan, place the prawns on their stomachs on top of the salt, in an alternate order of heads to tails, cover.

3. Pan-broil for 3 mins on medium heat till little steam appears, reduce to medium low heat for 3~5 mins till done (for electric griller, preheat to 400°F and pan-broil for 6~8 mins till done); check for doneness, serve with tomato and leafy greens.

Culinary Tips:

- Traditionally, prawns are usually marinated with baking soda or sodium borax to keep them firm; the prawns are then grilled directly or brushed with butter and baked in an oven; this results in the prawns being either too dry or too greasy; also the prawns will curl and shrink easily, losing their nutritious, natural flavor-filled taste.

- The legs are trimmed off completely so that the prawns can stand upright in the skillet; and pepper is sprinkled on the stomach so that the flavor will seep through the thin membrane.

- For high quality ingredients, pan-broiling is the best cooking method because it not only preserves the abundant flavor, nutrients, digestive enzymes and color, it enables you to savor the delicately smooth and elastic texture of the prawns; for variation of flavor, the prawns can be served with lemon chilly sauce.

奶香淡菜盅

8分鐘 / 4人份 / 1人份299卡

lled Creamy Mussels

8 mins / 4 servings / 299 cal per serving

144

材料：

淡菜16粒（約600克）、奶油15克、洋蔥末75克、鮮奶油150c.c.、鮮奶150c.c.、羅勒葉末1茶匙

調味料：

鹽少許

作法：

1. 淡菜洗淨，用剛好淹過材料的冰過濾水及1大匙海鹽泡數小時，完全吐砂濾淨後，再放入8吋平底煎鍋中，加蓋。

2. 取3/4公升的小調味鍋，預熱至水珠慢跑，加入奶油、洋蔥末一起以中小火爆香約1分鐘至金黃色，開蓋並用濕布擦鍋緣，入鮮奶油及鮮奶，一起煮開，入鹽拌勻備用。

3. 以中火煮淡菜約2分鐘至水氣冒出及淡菜口微開，熄火，起鍋放入保溫的湯碗中，倒入作法2調味，撒上羅勒葉末即成。

原味達人的烹調祕笈

- 傳統料理奶香淡菜，大多把淡菜放入滾水中汆燙，容易縮水並失去鮮甜度；而傳統白醬則是使用較多奶油增添香氣，熱量較高，又無法分別吃出海鮮的原味，及享受白醬的奶香味。

- 此道原味烹調的重點，是以無油乾烤法保持淡菜原始的風味，並且利用鍋子乾烤直接逼掉海腥味，煮到剛好熟時，肉質飽滿，鮮甜又美味。

- 此道食譜，可依個人的口味，搭配全麥雜糧麵包沾著食用。

Ingredients:

600g/16 mussels, 15g butter, 75g onion (minced), 150 c.c. whipping cream, 150 c.c. milk, 1 tsp basil leaves (minced)

Seasonings:

dash salt

Method:

1. Wash mussels, immersed in chilled filtered water with 1 Tbsp of sea salt for a few hours to release the sand within thoroughly and drain; place them in an 8-inch skillet and cover.

2. Preheat a 3/4-quart saucepan till water droplets "dance slowly"; add butter and onion, cover; pan-fry for 1 min on medium low heat till golden yellow; uncover, clean edge of pan with wet cloth, add whipping cream, then milk, bring to a boil, add salt and stir well.

3. Pan-broil mussels for 2 mins on medium heat till steam appears and they open up slightly; transfer into warm bowls and ladle creamy sauce prepared in step 2 over the mussels; sprinkle basil and serve immediately.

145

Culinary Tips:

- Traditionally, this dish is prepared by parboiling the mussels first, causing it to shrink and lose its natural sweetness; also the cream sauce is prepared with much butter, as well as adding excessive calories, to restore flavor.
- Using the greaseless and waterless cooking method, the mussels are pan-broiled to preserve the natural goodness while eliminating the unpleasant scent; the mussels is also naturally flavorful and juicy, greatly reducing food shrinkage when cooked with perfect timing.
- This butter cream sauce can be served as a dip with whole wheat bread.

西芹番茄炒墨魚 15分鐘 / 4人份 / 1人份141卡

Pan-Broiled Squid with Celery and Tomatoes Stir-Fry
15 mins / 4servings / 141 cal per serving

材料：

透抽1條（約500克）、西洋芹菜斜段250克、
紅蘿蔔片60克（0.5公分）、牛番茄塊300克、
甜椒片100克（2.5公分斜片）、紅辣椒段1根、
薑片4片、芥花油1茶匙

醃料：

蔥薑蒜酒汁1大匙

調味料：

鹽、黑胡椒各適量

作法：

1. 透抽用粗鹽及檸檬皮搓洗，再用過濾水沖淨，
   濾乾水分，加入醃料醃約30分鐘，下鍋前先
   擦乾。

2. 取12吋平底煎鍋，預熱至水珠快跑，鍋內一
   邊放入透抽貼鍋乾煎，撒入鹽；另一邊倒入
   油、薑片、西洋芹菜段、紅蘿蔔片，番茄、甜
   椒片及紅辣椒段鋪在西洋芹上面，加蓋。

3. 將鍋面偏離火中心（蔬菜邊的鍋面離火較
   遠），以中火煎約1分30秒，轉小火，開蓋，
   用濕布擦鍋緣，將透抽翻面並稍微拌炒其他的
   蔬菜，撒入鹽及黑胡椒，加蓋，以中火燜炒約
   1分30秒～2分鐘。食用前再將透抽切成圈狀
   即成。

原味達人的烹調祕笈

利用少油無水烹調，可以一鍋兩用，一邊無油煎
烤透抽，一邊少油炒蔬菜，讓食材同步煮熟。透
抽整隻入鍋不用油煎可去腥並避免出水；蔬菜只
用少許油及不加水的情況下，藉著受熱產生的蒸
汽煮熟，不加太多調味料就很鮮美。

料理時要特別注意火候，透抽用中火，蔬菜用中
小火，所以讓鍋子偏向一邊，使透抽受熱溫度較
高，表皮略呈金黃色，口感Q中帶嫩，特別鮮香
甜美，而蔬菜也不會燒過熟，保留原汁原味。

Ingredients:

500g/1 squid, 250g celery (segmented diagonally), 60g carrot (sliced, 0.5 cm), 300g/2 tomatoes (quartered), 100g bell peppers (2.5 cm sliced diagonally), 1 red chilly (sliced), 4 slices ginger, 1 tsp canola oil

Marinate:

1 Tbsp mixed juice of scallion, ginger, garlic and wine

Seasonings:

dash salt, dash coarse black pepper

Method:

1. Clean squid with lemon skin and sea salt, rinse and drain, marinate for 30 mins, pat dry before cooking.

2. Preheat a 12-inch skillet till water droplets "dance quickly"; place squid on one side of the skillet, sprinkle salt; on the other side, add oil, ginger, celery, carrots, and then tomatoes, bell peppers and chilly on top of the celery, cover.

3. Shift the skillet so that the side with veggies is farther away from the heat; pan-broil the squid and pan-fry veggies for 1.5 mins; turn to low heat, uncover, clean edge of pan with wet cloth, flip the squid, and stir-fry the vegetables, sprinkle salt and black pepper, cover; fry for 1.5 mins on medium heat, give a quick stir, cut up the squid into rings before serving.

147

Culinary Tips:

· By pan-broiling the squid in whole and cooking together with the veggies with a little oil and without adding water, preserves all the nutrients of every ingredient to the maximum, so only a small amount of seasoning needs to be added; also the squid is browned till fragrant and it is delicately sweet and chewy; the veggies when not overcooked is naturally flavor-filled.

· However, when cooking this recipe, be cautious of the heat applied; the squid needs to be cooked on medium heat, to preserve the natural juices, whereas the veggies to be cooked on medium low; therefore the skillet is positioned so that the heat is applied more directly on the squid.

橙汁軟絲

15分鐘 / 6人份 / 1人份113卡

Lemony Orange Glazed Squid

15 mins / 6 servings / 113 cal per serving

148

材料：

軟絲1條（約800克）、巴西里末少許、橙皮絲及檸檬皮絲適量

醃料：

蔥薑蒜酒汁1大匙

調味料：

鹽少許、新鮮柳橙汁1/4杯、檸檬汁1/4杯、楓糖漿2大匙、香菇醬油1/2茶匙

作法：

1. 軟絲去頭、剝皮再用粗鹽及檸檬皮洗淨並濾乾水分，不切開，加入蔥薑蒜酒汁醃約30分鐘，再吸乾水分。

2. 取12吋平底煎鍋，預熱至水珠快跑，放入軟絲仔貼緊鍋面，撒入少許鹽，加蓋以中火乾烤約3分鐘，轉小火，開蓋翻面以中小火烤約3分鐘。

3. 轉小火，取出軟絲，放入調味料，以中小火煮開至稍微收汁，再放入軟絲，將兩面均勻上色，撒入巴西里末、橙皮絲及檸檬皮絲即成。

原味達人的烹調祕笈

- 傳統料理軟絲會刻花切塊再直接放入滾水汆燙，再加入許多油及調味料拌炒勾芡。軟絲經過二次烹調料理的動作，食物會縮水吃不到其營養與原味。

- 無油烹調軟絲，等到整隻烤好再切塊，肉質的飽水度更高，吃起來彈Q滑嫩。

- 不用冰糖而改用楓糖漿做調味料，口感較不甜膩，同時在調味料中添加新鮮柳橙汁及檸檬汁，使味道微酸微甜，還有天然的果香。

Ingredients:

800g/1 neritic squid, dash English parsley (minced), dash orange zest, dash lemon zest

Marinade:

1 Tbsp mixed juice of scallion, ginger, garlic and wine

Seasonings:

dash salt, 1/4 cup fresh orange juice, 1/4 cup lemon juice, 2 Tbsp maple syrup, 1/2 tsp mushroom-flavored soy sauce

Method:

1. Peel off membrane, pull off head, discard unwanted materials from head; clean with lemon skin and sea salt, rinse and drain, marinate for 30 mins, pat dry before cooking.

2. Preheat a 12-inch skillet till water droplets "dance quickly"; place the squid flat in the skillet, sprinkle salt, cover; pan-broil for 3 mins on medium heat; turn to low heat, uncover, flip; pan-broil for 3 mins on medium low heat.

3. Turn to low heat, remove squid, add seasonings to deglaze the pan for sauce; turn the heat up to medium small till it liquidizes slightly; place the squid in the skillet again, glaze evenly; sprinkle parsley, orange zest and lemon zest.

Culinary Tips:

· Traditionally, multiple deep cuts are made on the inside surface of squid; they are prepared by parboiling first and then quick-fried in much oil; this 2-step cooking process destroys both the nutritional value and natural flavor and causes food shrinkage; thus starch is normally used to thicken the sauce to absorb the seasonings that is necessary to restore flavor.

· With greaseless cooking, when pan-broiled, the natural qualities are locked into food, keeping the aroma in the pan; the squid is only cut when served, leaving the squid juicy and tasty.

· Maple syrup instead of rock sugar is used; fresh orange and lemon juice is added to enhance the flavor; therefore it gives a light, refreshing and fruity taste.

鍋烤龍蝦 10分鐘 / 4人份 / 1人份468卡

Grilled Lobster

10 mins / 4 servings / 468 cal per serving

材料：

加拿大小龍蝦2隻（約1200克）、麻油1茶匙、
老薑8片

醃料：

蔥薑蒜酒汁2大匙

作法：

1. 小龍蝦用棕櫚刷洗淨，放入濾網濾乾水分，醃
   蔥薑蒜酒汁並放入冷凍庫1小時。
2. 取12吋平底煎鍋（或可用電烤鍋預熱至
   325℉），預熱至水珠慢跑，放入麻油、老薑
   片，以小火爆香兩面各30秒。
3. 再放入小龍蝦，薑片移至龍蝦上，加蓋，以中
   火烤約3分鐘至水氣冒出，轉中小火續烤5分鐘
   （可用電烤鍋轉至400℉烤8分鐘），用不鏽
   鋼刺針刺蝦腹最厚處，容易刺穿、肉質有彈性
   即成。

原味達人的烹調祕笈

- 活龍蝦在合宜的冷藏空間，可以無水存活1～2
  天；只要將龍蝦用濕潤的報紙或海草覆蓋再放入
  保麗龍盒，冰在冰箱即可。下鍋前，將龍蝦洗淨
  再放入冷凍庫1小時，使其慢慢凍死，才不會因
  入熱鍋時受驚嚇肉質變硬；或用刀子刺蝦頭連接
  蝦身的中間，昏死再入鍋。
- 傳統料理多是入滾水汆燙，或剖半加醬料以大火
  蒸，肉質縮水，口感易乾硬。
- 龍蝦貼鍋乾烤可增添香氣，搭配麻油炒老薑可以
  驅寒，煮至剛好的熟度，即能享受到龍蝦肉質彈
  性的Q度及鮮美滋味。
- 食用龍蝦時，先取專用的剪刀從兩邊剪開蝦腹的
  薄殼，即可輕鬆取出蝦肉，再用湯匙舀取蝦膏；
  吃剩的蝦殼或蝦腳，可加入滾水熬成龍蝦高湯。

Ingredients:

1,200g/2 small Canadian lobsters, 1 tsp sesame oil,
8 slices aged ginger

Marinade:

2 Tbsp mixed juice of scallion, ginger, garlic and wine

Method:

1. Clean lobsters with palm brush, rinse, drain,
   marinade and put in freezer for 1 hour.
2. Preheat a 12-inch skillet (for electric griller,
   preheat to 325℉) till water droplets "dance
   slowly"; add sesame oil and ginger, fry both
   sides for 1/2 min on low heat.
3. Place lobsters right-side-up on pan in an alternate
   order of head to tail, put ginger on them, cover;
   pan-grille for 3 mins on medium heat till steam
   appears, turn to medium low heat for 5 mins till
   done; (for electric griller, turn to 400℉, pan-
   grille for 8 mins); wipe lid with wet cloth, check
   for doneness with cooking pin by piercing the
   thickest part of stomach, it should be easily
   pricked and have an elastic texture.

Culinary Tips:

- Live lobster can live out of water for 1 to 2 days
  under ideal cool, damp storage conditions. They
  can be stored in the fridge in a foam box covered
  with damp newspaper (salt water) or seaweed.
  The most humane method to prepare live lobster
  is to place lobster in the freezer for 1 hour. It
  will not freeze the meat but has an anaesthetizing
  effect on the lobster. Alternatively to kill lobster
  instantly, pierce the shell at the centre of body,
  behind the head. Never cook live lobster instantly
  as this toughens the meat.
- Traditionally, it is prepared by boiling in water or
  cutting into halves and steam with sauce. The
  meat will tend to shrink and taste dry and stale.
- Adding sesame oil and ginger warms your body.
  By pan-grilling without water and not overcooked,
  the delicate flavor of its natural freshness and the
  elastic texture is enough to make this dish very
  palatable.
- When eating lobster, first cut the thin abdominal
  shell with a pair of scissors to extract the lobster
  meat. Use a spoon to enjoy the creamy paste
  inside the lobster's head. The shell or extras of
  the lobster can be boiled and become a base for
  soup.

151

鮮味蝦 10分鐘 / 3人份 / 1人份113卡

10 mins / 3 servings / 113 cal per serving

材料：

鮮活海鱸蝦300克（約12尾）、蔥段2根、薑5片、蒜頭2粒

調味料：

米酒1.5茶匙

作法：

1. 取不鏽鋼調理盆，加入少許冰塊再放入鱸蝦，蝦頭不易變黑；用筷子攪動，可輕易讓鮮蝦吐砂又不傷手。

2. 修剪蝦鬚、蝦頭尖、蝦腳，用牙籤挑除腸泥，洗淨並濾乾，放入8吋平底煎鍋中，放入其它輕拍過的材料，加蓋，直接放入冰箱冷藏室保鮮。

3. 食用前取出，濾乾多餘的水分，入調味料，加蓋再整鍋放在爐上，以中火煮3分鐘至鍋蓋有水氣冒出來，輕輕搖鍋再繼續煮1分鐘，熄火即取出盛盤。

原味達人的烹調祕笈

※ 鮮蝦最常見的傳統料理，大多是高溫油炸、滾水汆燙或大火蒸煮等方式，需添加很多佐料或調味料提味，容易流失營養又吃不到鮮蝦的鮮甜原味。

※ 鮮蝦處理完成後，直接放入不鏽鋼鍋中，加蓋移入冰箱冷藏，可繼續保持新鮮度；端到火爐上烹調，利用食物本身的熱氣循環原理，瞬間燴熟，即可吃到自然鮮甜、原味又彈牙的蝦料理。

Ingredients:

300g /12 live ocean "lu" shrimps, 2 scallions (segmented), 5 slices ginger, 2 garlic (crushed)

Seasonings:

1.5 tsp rice wine

Method:

1. In a stainless steel container, place shrimps on ice, to preserve the heads from turning black easily; stir with a pair of chopsticks to allow shrimps to release the sand while not injuring one's hands.

2. Trim off whiskers, sharp head and legs of the shrimps, devein with a toothpick, rinse clean and drain. Place shrimps in an 8-inch skillet, add all the other ingredients that were crushed to season, cover and chill in the fridge.

3. Before serving, drain off excess liquid, add seasonings, cover; place skillet on stove and baste in its vapor steam for 3 mins on medium heat, till steams appears, shaking the skillet slightly after 1 min; wipe lid with wet cloth, check for doneness and serve.

Culinary Tips:

· Traditionally, the most common method of cooking shrimps is to stir-fry with much oil, deep-fried, boil or steam, adding much seasonings for more flavor. However, it tends to lose its natural nutrients and flavor.

· By chilling the shrimps in the stainless steel skillet to better preserve the freshness, and moving directly on the stove to baste in its own vapor steam till just cook, will allow you to taste the natural sweetness and delicate flavor plus the elastic texture of the prawn.

南瓜薯泥沙拉 30分鐘 / 4人份 / 1人份145卡

Mixed Pumpkin Potato Salad

30 mins / 4 servings / 145 cal per serving

154

材料：

小馬鈴薯200克（約3顆）、南瓜200克（約1/8顆）、紅蘿蔔40克（1/4條）、嫩玉米100克（約1/2根）、雞蛋1顆、蘋果丁75克（約1/2顆）、小黃瓜丁50克、明日葉絲10克、蔓越莓適量

調味料：

檸檬優格醬3大匙、鹽1/4茶匙、檸檬汁1大匙

作法：

1. 馬鈴薯、南瓜、紅蘿蔔、玉米洗淨，與雞蛋一起放入1公升鍋中，倒入過濾水1/2杯以中火煮沸，轉小火續煮約3分鐘，取出玉米、雞蛋待涼，轉中火煮開再轉小火。玉米剝成顆粒狀，剝除蛋殼，並預留少許蛋黃末。

2. 經過3分鐘後，取出南瓜；9分鐘後取出馬鈴薯及紅蘿蔔，去皮。南瓜及馬鈴薯趁熱壓成泥狀，胡蘿蔔切成小丁，再放入檸檬優格醬、鹽拌勻。

3. 將蘋果丁、小黃瓜丁與檸檬汁一起拌勻，再放入作法2的沙拉泥中拌勻，上面搭配蔓越莓、明日葉絲、蛋黃末即成。

原味達人的烹調祕笈

- 沙拉要保持原味口感，須改良傳統的料理方式，例如帶皮的食材不需脫皮煮熟；每種食材要以大塊烹煮，還要煮到剛好透的熟度以維持原味；以少水烹煮取代大量滾水；不添加高熱量的美乃滋，改用自製的天然佐料調味。

- 以無水烹調法料理帶皮的沙拉食材，可輕易去除表皮，更能鎖住食材的營養及甜分，特別是馬鈴薯及南瓜泥，不需靠美乃滋就有自然的黏稠度。

- 多種不同屬性的食材同時入鍋烹調時，要鎖住營養到最高點，需依先後順序取出，且每次取出須馬上加蓋並重新轉中火，讓水氣釋放出來，再開始計時。

Ingredients:
200g/3 small potatoes, 200g/1/8 pumpkin, 40g/1/4 carrot, 100g/1/2 corn, 1 egg, 75g/1/2 apple (diced), 50g cucumber (diced), 10g ashitaba (shredded), dash cranberries

Seasonings:
3 Tbsp lemon yogurt sauce, 1/4 tsp salt, 1 Tbsp lemon juice

Method:

1. Wash potatoes, pumpkin, carrot and corn; place them with the egg into a 1-quart pan, add 1/2 cup filtered water, bring to a boil on medium heat; turn to low heat for 3 mins, remove corn and egg to cool; turn to medium heat till steam appears and then adjust to low heat; peel corn into nib lets, shell the egg, reserving some minced yolk.

2. Remove pumpkin after 3 mins, potato and carrot after another 9 mins and peel; while still warm, mash pumpkin and potato, dice carrot, add lemon yogurt sauce and salt; mix well.

3. Mix diced apple and cucumber with lemon juice, add to the salad prepared in Step 2; top with cranberries, ashitaba and yolk.

Culinary Tips:
- In order to preserve the natural goodness, there is a need to improve the traditional methods, such as cooking ingredients in whole or in big cuts with minimum moisture (instead of boiling and immersed), without peeling and over-cooking; eliminating the need for added calories with much mayonnaise, but rather using the healthy homemade dressings.
- By cooking the ingredients through "waterless cooking", besides preserving the maximum nutrients, the potato is easily peeled and especially the texture of the potato and pumpkin is stickier and thicker so there is no need for mayonnaise.
- When cooking different ingredients with varying nature together, they should be removed in sequence according to their cooking times; each time to remove, make sure to cover, turn to medium heat until steam appears again and reset the timer.

苦盡甘來

20分鐘 / 4人份 / 1人份191卡

20 mins / 4 servings / 191 cal per serving

材料：

綠苦瓜1條（約500克）、香菜末及烤酥的白芝麻各適量、芥花油1茶匙

調味料：

碎冰糖1大匙、香菇醬油1/4杯

作法：

1. 綠苦瓜用軟刷充分洗淨，切成3公分厚的圓圈狀（不要去籽）。
2. 取8吋平底煎鍋，預熱至水珠慢跑（或鐵板電烤鍋轉350℉），放入芥花油並抹勻，擺入苦瓜圈，以中小火煎兩面各約3分鐘呈金黃色。
3. 將碎冰糖放入苦瓜中心，均勻淋入香菇醬油及水1大匙，加蓋以中火煮沸，轉小火（或鐵板電烤鍋轉250℉），兩面燜煮各約5分鐘至收汁熟透，撒入白芝麻及香菜段即成。

原味達人的烹調祕笈

- 傳統作法大多先將苦瓜去籽切塊，放入熱油鍋過油去苦味，再加入大量的水、醬油及冰糖以小火慢慢燒煮至收汁，這種作法容易流失食材的養分，吃不到原來的苦瓜味。
- 苦瓜要保留原味的作法非常簡單，切大塊的圈狀、保留苦瓜籽、以少油清煎兩面至八分熟，可去苦味，增加香氣，再以少量的醬汁燒煮。富含維生素E及酵素的苦瓜籽，吃起來別有風味。

Ingredients:

500g/1 green bitter gourd, dash parsley (minced), dash pan-grilled white sesame, 1 tsp canola oil

Seasonings:

1 Tbsp rock sugar (crushed), 1/4 cup mushroom-flavored soy sauce

Method:

1. Scrub clean bitter gourd with a soft brush, cut into rounds, 3 cm thick; do not discard the seeds.
2. Preheat an 8-inch skillet till water droplets "dance slowly" (for electric griller, preheat to 350℉); add oil, coat evenly, place the bitter gourd rounds, pan-grille for 3 mins each side same for griller on medium low heat till golden brown.
3. Fill the center of the rounds evenly with crushed rock sugar, pour on soy sauce and 1 Tbsp of water, cover and bring to a boil on medium heat; simmer each side for about 5 mins on low heat (for electric griller, turn to 250℉), till the sauce liquidize; wipe lid with wet cloth, test for doneness, sprinkle with white sesame seeds and minced parsley; serve.

Culinary Tips:

- Bitter gourd is traditionally prepared by discarding the seeds and cutting into pieces; it is then deep-fried to eliminate its bitterness; much soy sauce, water and rock sugar are added and then liquidized; this method tends to lose its nutritional value and natural flavor.
- It is easy to preserve the natural goodness of the bitter gourd, cut into large rounds, preserve the seeds, pan-grille in minimum oil till medium well done and golden brown, to eliminate the bitterness while increasing its fragrance and then cooking in little sauce; the bitter gourd's seeds, rich in Vitamin E and enzymes makes this dish tastes special and have a distinctive flavor.

滑蛋絲瓜

10分鐘 / 3人份 / 1人份59卡

10 mins / 3 servings / 59 cal per serving

材料：

絲瓜1條（約500克）、雞蛋1顆、蔥白段1根、嫩薑絲適量、芥花油1.5茶匙

調味料：

鹽1/4茶匙

作法：

1. 絲瓜洗淨，去薄皮後，切成四長條再斜切成約2公分寬的菱形塊狀，下鍋前用水潤濕再濾乾；雞蛋打散拌勻。

2. 取8吋平底煎鍋，預熱至水珠慢跑，加入芥花油，放入蔥白段、嫩薑絲以中小火爆香半分鐘，轉小火，續入絲瓜（綠皮肉朝下），蔥白段及薑絲移至絲瓜上，加蓋以中小火炒兩面各2分鐘。

3. 轉小火翻面，用濕布擦鍋緣，再加蓋以中火繼續炒約1～2分鐘至熟透，加入鹽拌勻，均勻淋入蛋液，加蓋，熄火燜煮約30秒即成。

Ingredients:

500g/1 loofah gourd, 1 egg, 1 scallion (segmented), dash young ginger (shredded), 1.5 tsp canola oil

Seasonings:

1/4 tsp salt

Method:

1. Wash clean, lightly peel skin; cut into quarters horizontally and then diagonally into 2 cm wide diamond shape chunks, rinse and drain; whisk the egg.

2. Preheat an 8-inch skillet till water droplets "dance slowly"; add oil, pan-fry scallion and ginger for 1/2 min on medium low heat; turn to low heat, add loofah (skin side down), move scallion and ginger on top, cover; simmer for 2 mins each on 2 sides over medium low heat.

3. Turn to low heat, clean edge of pan with wet cloth, flip, cover; turn to medium heat and cook for 1-2 mins till the loofah is just cooked; stir in the salt, pour the beaten egg evenly in circles, cover; turn off the heat and simmer for 1/2 min before serving.

原味達人的烹調祕笈

傳統料理絲瓜時，最常作法是煮絲瓜加水，那是因為絲瓜切太小塊或太薄，造成絲瓜白色的果肉吸收了鍋中的油質而容易燒焦，只好加水烹調，這樣反而會流失其本身的鮮甜味。

絲瓜是含水量多的食材，切成大塊菱形狀煮熟後，可保留較多甜分，還有料理時，將絲瓜綠色肉外皮貼著鍋底，可避免白色海綿組織吸油。

Culinary Tips:

· Traditionally loofah is cut into smaller pieces and tends to absorb all the oil in the pan and burn the food; therefore, much water is added and causes it to lose its natural sweetness.

· Loofah comes naturally with much moisture; by cutting the loofah into large diamond chunks, it is more naturally sweet and nutritious; pan-frying the skin layer first, prevents the inner layer to soak up the oil content instantly and allows food to be cooked in its own natural moisture when vapor escapes around the lid.

159

胡瓜蝦仁炒蛋

10分鐘 / 4人份 / 1人份137卡

Big Cucumber with Shrimp Omelet

10 mins / 4 servings / 137 cal per serving

材料：

鮮蝦仁120克、大黃瓜1條（約500克）、紅蘿蔔片60克（厚0.5公分）、雞蛋3顆、紅辣椒片1根、薑片3片、蒜末1茶匙、芥花油1.5茶匙

調味料：

鹽1/2茶匙、白胡椒粉少許

作法：

1. 將大黃瓜用海棉刷洗乾淨，直切4等分後去籽，再斜切片成約1公分厚；蝦仁去腸泥，洗淨；雞蛋打散，入1/4茶匙鹽拌勻。

2. 取12吋平底煎鍋，預熱至水珠慢跑，加入蒜末、芥花油，以小火爆香1分鐘，取出備用；再抹少許油，在鍋內2/3處依序放入蝦仁、大黃瓜及紅蘿蔔片，1/3處則倒入蛋液，均勻撒入爆香好的蒜末，加蓋，以中火煎約1分鐘。

3. 用濕布擦鍋緣再開蓋，均勻撒入少許白胡椒粉及1/4茶匙的鹽，再將大黃瓜、紅蘿蔔片及蝦仁直接用平鏟盛起置於半熟的蛋液上，再快速用鍋鏟，從蛋的底部翻面，加蓋，貼鍋繼續以中火煎約1分鐘即成。

原味達人的烹調祕笈

- 蔬菜炒蝦仁在傳統料理，一般都是把大黃瓜切細；蝦仁過油或可能加硼砂增加脆度；用熱油炒蛋，分開拌炒的結果就是加更多次的油，不但較油膩又無法吃到每樣食材的原味。

- 我們利用一個大煎鍋，一氣呵成完成這道菜，只有加一次油，並且將平底煎鍋分成三等分，放入不同的食材，不需翻炒即可保持食材的原味，又可以讓蝦仁貼鍋去腥，口感較香Q又清甜。

Ingredients:

120g fresh shrimps (shelled), 500g/1 big cucumber, 60g carrots (sliced, 0.5 cm), 3 eggs, 1 red chilly (sliced), 3 slices ginger, 1 tsp garlic (minced), 1.5 tsp canola oil

Seasonings:

1/2 tsp salt, dash white pepper

Method:

1. Scrub clean cucumber, with a sponge; cut into quarters horizontally, discard seeds, cut diagonally into 1 cm thick slices; de-vein and wash shrimps; whisk eggs and add 1/4 tsp salt.

2. Preheat a 12-inch skillet till water droplets "dance slowly"; add garlic, oil, fry for 1 min on low heat till golden yellow, remove; spread oil evenly, lay out shrimps, cucumber, carrots on 2/3 of the pan, pour the eggs on 1/3 pan and sprinkle fragrant garlic on shrimps, cover; pan-fry for 1 min on medium heat.

3. Uncover, clean edge of pan with wet cloth, add dash pepper and 1/4 tsp salt on the cucumbers, carrots and shrimps, scrape them onto the partially cooked eggs; using a slice, quickly flip the eggs over, cover; pan-fry for another min on medium heat till cooked, serve.

Culinary Tips:

- Traditionally, this dish is much greasier and against the concept of natural cooking because each ingredient is cooked individually in hot oil; also the ingredients are cut smaller and the shrimps are scalded in oil or seasoned with sodium borate to make them crisp.

- Using a larger skillet to cook all the ingredients at the same time, only add oil once in the process and the food are cooked in its own natural moisture without frying and stirring, to lock in the nutrients; pan-frying the shrimps, covered, can also eliminate its unpleasant seafood scent.

161

魚香茄子

15分鐘 / 4人份 / 1人份179卡

Eggplant Topping with Ground Pork

15 mins / 4 servings / 179 cal per serving

162

材料：

茄子400克、瘦豬絞肉150克、肥豬絞肉10克、蒜末1茶匙、薑末1茶匙、紅辣椒末1根（約15克）、蔥末10克、九層塔末適量、芥花油1茶匙

調味料：

香菇醬油1大匙、米酒1/2茶匙、白胡椒粉1/4茶匙、紅蔥酥1茶匙

作法：

1. 取1公升鍋，預熱至水珠快跑，放入肥豬絞肉，以中火乾爆兩面各30秒至油逼出，轉小火，放蒜末、薑末爆香至金黃取出，再續入瘦豬絞肉抹平，放回爆香的蒜末及薑末，加蓋約1分鐘至小水氣冒出，打開鍋蓋拌勻。

2. 茄子洗淨，去除蒂頭，切6公分長段，泡鹽水1分鐘。取11吋平底煎鍋，預熱至水珠慢跑，入芥花油、茄子，加蓋，以中火煎約2分鐘至深紫色，轉小火，翻面，再轉中小火續煎約2分半鐘（可用食物刺針試熟度，若不夠透再翻第三面以中小火煎約1～2分鐘）。

3. 加調味料拌勻，加蓋以中小火煮至水氣出，轉小火燜煮約2分鐘，再放辣椒末、蔥末及九層塔末拌炒，起鍋鋪在煎熟的茄子上即成。

原味達人的烹調祕笈

- 傳統作法將茄子切滾刀塊或薄片狀，剖面較大，入鍋油炸，容易吃油又需加水燜煮，不健康又喪失原味。

- 將茄子直切成長段並保留外皮，可減少剖面避免接觸空氣變色，又能阻隔海綿體的果肉吸收多餘油分。只需少許油清煎至外皮呈深紫色，靠著烹煮時產生的蒸汽燻熟茄肉，吃起來多汁、入口即化又不油膩。

- 一般餐廳作法使用豬絞肉的比例（肥油：瘦肉＝3：7）混著炒，造成油量要多，如果油量不多，瘦豬絞肉易出水變得不香。利用乾爆肥肉末所逼出的油脂來爆香全瘦豬絞肉，不需額外加油還可減少肥油比例。

Ingredients:

400g eggplants (segmented, 6 cm), 150g minced lean pork, 10g minced fatty pork, 1 tsp garlic (minced), 1 tsp ginger (minced), 15g/1 red chilly (minced), 10g scallion (minced), dash basil (minced), 1 tsp canola oil

Seasonings:

1 Tbsp mushroom-flavored soy sauce, 1/2 tsp rice wine, 1/4 tsp white pepper, 1 tsp fried shallots

Method:

1. Preheat a 1-quart pan till water droplets "dance quickly"; spread fatty pork on pan, pan-broil for 1/2 min each, on both sides over medium heat till fat is released; turn to low heat, add garlic and ginger, pan-fry till golden yellow, then remove; spread lean pork, the fragrant garlic and ginger, cover; pan-fry for 1 min each, on both sides over medium heat till little steam appears.

2. Soak eggplants in salt water for 1 min, rinse; preheat an 11-inch skillet till water droplets "dance slowly"; add oil, eggplants, cover; pan-fry for 2 mins on medium heat till skin turns dark purple, turn to low heat, flip; pan-fry for 2.5 mins on medium low heat, test for doneness (if eggplant is thick, flip on 3rd side to fry for another 1 to 2 mins on medium low heat till done).

3. Add seasonings to step 1, stir, cover; cook on medium heat till little steam appears, turn to low heat; simmer for 2 mins, garnish with red chilly, scallion and basil and stir; dish onto the cooked eggplant and serve.

163

Culinary Tips:

- Traditionally, eggplant is being cut into hop pieces or thin slices; if deep-fried, the larger cut section will easily absorb more oil and is simmered with much water, losing its nutrition.

- Cutting into long segments will limit the area exposed to air becoming dark, also only minimum oil is necessary to pan-fry the eggplant; basting in its own vapor steam preserves the rich, appealing and tender melt-in-your mouth texture.

- Only a little fatty pork which is bought separately, is used is this recipe; the fat is released to brown the garlic, ginger and lean pork; whereas in most restaurants, the ratio between fatty and lean pork is 3:7.

鍋烤栗子

35分鐘 / 8人份 / 1人份185卡

Pan-Grilled Chestnuts

35 mins / 8 servings / 185 cal per serving

材料：

新鮮的天津栗子800克

作法：

1. 栗子先用水洗淨，再用濾網濾乾水分。
2. 取12吋鐵板電烤鍋，預熱至400°F，再放入栗子，加蓋，烤約18分鐘，中途每隔5分鐘要左右晃動鍋子。
3. 鐵板電烤鍋轉325°F，續烤約18分鐘至表面出油，起鍋前試吃栗子是否有鬆軟帶Q的口感即成。

原味達人的烹調祕笈

- 傳統習慣用烤箱烤栗子，外殼易爆開破裂，果肉也較乾；或將栗子加水煮至半熟，再放入烤箱烤熟，同樣失去香味；而市售的烤栗子通常使用砂石粒加入糖，不停滾動翻炒，因翻炒空間大，較易流失甜分及香氣。

- 栗子要保留原味，最好使用氣密性佳的鐵板電烤鍋恆溫乾烤，先用400°F高溫，讓它產生香氣，經18分鐘後，再調降至325°F保持水分，吃起來比傳統糖炒栗子更香甜，還能鎖住水分，口感又鬆又香。

- 栗子要趁熱食用，口感較鬆，如果冷掉會繼續熟成，雖有甜味，但水分會稍乾。

Ingredients
800g fresh Tianjin chestnuts

Method:
1. Wash clean, drain the chestnuts.
2. Preheat 12-inch electric griller to 400°F, spread chestnuts evenly, cover; pan-grille for 18 mins, shaking the griller to and fro, covered, every 5 mins during the whole process.
3. Adjust electric griller to 325°F for another 16 to18 mins; test for doneness, shell chestnuts, the meat should give a slight shiny greasiness, taste if they are flavor-filled with its natural sweetness with a slightly loose texture and serve.

Culinary Tips:
- Traditionally, chestnuts are roasted in an oven, the shell cracks easily and taste drier; or before roasting, they are boiled until half cooked, losing their natural flavor and aroma; on the market, they are rotated in a large burner with pebbles and sugar.
- Initially pan-grilling the chestnuts at 400°F to emerge its fragrance, thereafter 325°F to retain its moisture, enables this easy-to-cook snack tastier than the traditional ones; it is best on griller due to its constant heat.
- Chestnuts should be eaten while still hot; the retained heat will overcook them and taste drier, even though still sweet.

芋頭南瓜派

35分鐘 / 12人份 / 1人份300卡

Pumpkin Pie

35 mins / 12 servings / 300 cal per serving

材料：

檳榔芋頭1顆（約800克）、帶皮黃色南瓜400克、七葉蘭6葉、新鮮帶殼白果200克、薄荷葉少許、芥花油4茶匙

調味料：

砂糖165克、鹽1茶匙、楓糖2大匙

作法：

1. 芋頭去皮，放入2公升鍋中，加入過濾水1杯及七葉蘭，以中火煮開，轉中小火煮2分鐘再轉小火煮約20分鐘至熟透，熄火趁熱壓成泥狀，加入砂糖150克、鹽1小匙及芥花油3茶匙，以小火邊煮邊攪動至滑Q，即成芋泥。

2. 南瓜帶皮去籽放入1公升鍋中，加過濾水1/2杯，以中火煮開再轉小火煮約6～8分鐘至熟，濾掉煮剩的水分，趁熱壓成泥，加砂糖15克、芥花油1茶匙，轉中小火邊煮邊攪動至光滑狀，即成南瓜泥。

3. 取8吋平底煎鍋，預熱至水珠慢跑，放入白果，以小火烘烤約8～10分鐘（每3分鐘輕輕搖鍋），剝除外殼，加入楓糖漿燉煮至光亮黏稠狀。取一個模型，鋪上一層塑膠袋，先平均擺入白果，再放一層芋泥後，中間夾上南瓜泥，再添加一層芋泥，壓緊，脫模，裝入盤中，搭配薄荷葉即成。

原味達人的烹調祕笈

- 傳統料理芋頭南瓜派，會將食材切片狀，再入電鍋蒸，易流失原有的香氣和特殊的黏液，而需添加許多油及糖使其產生油光及黏滑口感。

- 南瓜及芋頭整顆煮，更能保留其黏液和甜分。芋頭加七葉蘭，另添清香。白果多是已處理好的真空包裝，但比不上手工烘烤的透明和香Q口感。

- 將剩餘的芋頭泥及南瓜泥，塗抹在新鮮的全麥潤餅皮上包成半圓形狀，放入溫度325℉的電烤鍋中，以少油煎3分鐘至兩面呈金黃色，切成小片狀，即可變化成好吃的南瓜芋泥煎餅。

Ingredients:

800g/1 taro (peeled), 400g/1 pumpkin (seeds discarded), 6 pine leaves, 200g fresh gingko nuts (with shell), dash mint leaves, 4 tsp canola oil

Seasonings:

165g cane sugar, 1 tsp salt, 2 Tbsp maple syrup

Method:

1. Place taro, pine leaves, 1 cup filtered water into 2-quart pot; bring to a boil on medium heat, simmer for 2 mins on medium low heat and then 20 mins on low heat till done; turn off heat, mash while still hot; add 150g sugar, 1 tsp salt, 3 tsp oil, stir on low heat till smooth and sticky.

2. Place pumpkin, 1/2 cup water into 1-quart pan; bring to a boil on medium heat, simmer for 6 to 8 mins on low heat till done; turn off heat, drain, mash while still hot; add 15g sugar, 1 tsp oil, stir on medium low heat till smooth.

3. Preheat 8-inch pan till water droplets "dance slowly"; pan-grille gingko nuts for 8 to10 mins on low heat, shaking the pan slightly every 3 mins; shell nuts, add syrup, simmer till shiny and thick; line a mold with plastic, layer with gingko nuts on the bottom, and taro and pumpkin paste alternately, press firmly, demould, garnish with mint leaves.

Culinary Tips:

- Traditionally, taro are sliced and steamed which makes them less sticky and tasty; therefore, much oil and sugar is added to enhance color and texture.

- Cooking taro as a whole, retains its natural sweetness and moisture; adding pine leaves makes it more flavorful. Gingko nuts are usually sold ready and packaged, but none can compare to pan-grilled nuts which is natural-flavored, transparent and chewy.

- The extra mixture can be spread on a fresh whole wheat skin (for Chinese spring roll), folded in semi-circle and pan-fry for 3 mins, each on both sides, with a little oil in an electric griller, preheat to 325℉ till golden brown.

冰鎮柴魚菠菜

5分鐘 / 2人份 / 1人份95卡

5 mins / 2 servings / 95 cal per serving

材料：

菠菜300克（約3棵）、烤酥的白芝麻1茶匙、柴魚片5克

調味料：

鹽及芝麻醬少許

作法：

1. 取1公升的調味鍋，預熱至水珠慢跑，放入柴魚片，加蓋以小火烤約1分鐘至酥，熄火取出。

2. 下鍋前，將菠菜潤濕並濾乾，再鬆鬆散開鋪放在11吋平底煎鍋，撒少許鹽，加蓋，以中火煮約1分鐘至小水氣冒出，熄火以濕布擦鍋蓋，開蓋將菠菜翻面，用餘溫燙熟，起鍋，放在冰塊上面冰鎮至涼。

3. 取出菠菜，擠出水分略乾，切成6公分長段，裝入盤中，撒入柴魚酥及白芝麻，再配搭芝麻醬料即成。

原味達人的烹調祕笈

- 菠菜傳統料理大多是先切段放入滾水中汆燙再漂冰水，而造成大量的營養素流失，吃起來的口感較澀，往往必須淋油使其滑順，而無法吃到菠菜本身的原味。

- 菠菜要煮出原味，最主要是利用菠菜下鍋前用水潤濕再濾乾時，表面附著的水珠，直接放入鍋中加蓋，利用食物本身在烹煮時產生的蒸汽煮透，保持蔬菜完整的原味及營養，口感清脆又不澀。

Ingredients:

300g spinach, 1 tsp white sesame seeds (pan-grilled), 5g dried bonito flakes

Seasonings:

dash salt, dash sesame sauce

Method:

1. Preheat 1-quart saucepan till water droplets "dance slowly"; add bonito flakes, cover; pan-grille for 1 min on low heat till crisp, remove.

2. Rinse and drain spinach before cooking; spread whole spinach loosely in an 11-inch skillet, sprinkle with salt, cover; pan-boil on medium heat for 1 min till little steam appears, turn off the heat; clean lid with wet cloth, overturn the spinach to cook thoroughly using the pan's retained heat, remove and chill on ice.

3. Drain spinach immediately till slightly dry, cut into segments, 6 cm long; garnish with bonito flakes and sesame seeds, serve with sesame sauce.

Culinary Tips:

· Traditionally, spinach is cut into segments, boiled by immersing in water, then blanch into iced water; it greatly reduces its nutrients and quality, resulting it to taste puckery and the need to drizzle with oil.

· Rinsing the spinach, then drain, allows enough water to cling to the veggies, to combine with its natural juices to cook in a bath of vapor steam which allows the nutrients to be instantly locked in the food, retaining its crisp, tender and smooth texture.

脆炒皇宮苗

5分鐘 / 2人份 / 1人份69卡

5 mins / 2 servings / 69 cal per serving

材料：

皇宮菜300克（約1把）、蒜頭3粒、芥花油1
大匙

調味料：

鹽1/8茶匙

作法：

1. 皇宮菜洗淨，下鍋前用水潤濕，用手摘成4
   公分段狀，葉與梗分開。
2. 取11吋煎鍋，預熱至水珠慢跑，加入拍扁的
   蒜頭及芥花油，加蓋，以小火爆香兩面各30
   秒至金黃色及香味散出。
3. 放入菜梗，再將菜葉舖在上面，加蓋，以中
   火炒約1分鐘至小水氣冒出來，用濕布擦鍋
   蓋再撒入鹽，拌炒均勻即成。

原味達人的烹調祕笈

炒青菜在傳統觀念裡，習慣猛火、熱油、加
水及開蓋快速翻炒，利用大量的雞粉或味精
來調味，與健康烹調的觀念完全背道而馳。
因為不斷翻炒的過程中，蔬菜吃進更多油；
高溫猛火會破壞蔬菜的維生素及礦物質而縮
水；開蓋炒食容易讓食物氧化變質；添加不
天然的調味料如味精會破壞味蕾。

蔬菜的原味料理，是利用蔬菜剛清洗好，表
面仍有的水珠烹調，將慢熟的菜梗先放入鍋
中，再把快熟的蔬菜葉擺在上面，加蓋，利
用蔬菜冒出的水蒸汽瞬間悶熟，吃起來清脆
又滑口，攝取更多的營養素。

Ingredients:

300g/1 bunch "Huang Kong" Spinach, 3 garlic (crushed), 1Tbsp canola oil

Seasonings:

1/8 tsp salt

Method:

1. Rinse and drain spinach before cooking; tear into 4 cm segments by hand, separating the leaves from stems.
2. Preheat an 11-inch skillet till the water droplets "dance slowly"; add garlic, oil, cover; and pan-fry for 1/2 min each, on both sides on low heat till golden yellow and allow the aroma to emerge; remove.
3. Place the stems in, the leaves and garlic on top, cover; pan-fry for about 1 min on medium heat till little steam appears; wipe lid with wet cloth, sprinkle with salt, stir till thoroughly cooked and serve immediately.

171

Culinary Tips:

· Traditionally, leafy vegetables are stir-fried constantly with much hot oil, under high heat and open cooking; also water and MSG is added to enhance the flavor; all these are against the concept of healthy cooking; firstly during constant stir-frying, the veggies absorb more oil; secondly, the high temperature destroys vitamins and minerals causing shrinkage; thirdly, in open cooking, oxidation of food occurs when food is exposed to the air, greatly reducing the food quality; fourthly, adding unnatural spices like MSG is unhealthy and it spoils our taste bud.

· Using the "waterless cooking" method, the stem which takes slightly longer time to cook is placed on the bottom, the leaves on top, with the lid on, to allow the veggies to be basted in a bath of vapor steam, giving the veggies a crisp and smooth taste and retaining its natural juices.

芥藍炒蘑菇 8分鐘 / 2人份 / 1人份87卡

8 mins / 2 servings / 87 cal per serving

172

材料：

芥藍菜300克、大蘑菇120克、薑片4片、紅蘿蔔片60克（厚0.5公分）、芥花油1茶匙

調味料：

鹽1/4茶匙

作法：

1. 洗淨的芥藍菜，烹煮前用水潤濕再切成4公分長段，葉與梗分開（厚梗則要斜切）；大蘑菇用過濾水洗淨，濾乾，切1公分厚片。

2. 取11吋平底煎鍋，預熱至水珠慢跑，加入薑片及芥花油，加蓋，以小火爆香兩面各1分鐘後，撥一邊，均勻抹油再分邊放入蘑菇片、紅蘿蔔片及菜梗，加蓋轉中小火炒約1分鐘。

3. 轉小火，打開鍋蓋，用濕布擦鍋緣再放入菜葉，撒鹽加蓋，轉中火炒約50秒至小水氣冒出來，開鍋蓋，快速拌炒即成。

原味達人的烹調祕笈

- 此道傳統作法，因芥藍菜帶有些苦味，梗不易熟，所以通常都會放入滾水中氽燙半熟，再加油拌炒，較會流失養分而需添加味精、鹽或蠔油來調味，也較不健康。

- 原味烹調強調每種食物都要煮出原味，所以這道菜要依食材部位的熟度屬性，調整放入鍋中烹調的順序，菜梗較慢熟先放進去，而葉片較快熟的最後放，才能達到同步煮熟，鎖住營養到最高點。

Ingredients:

300g cabbage mustard, 120g large button mushrooms (sliced 1 cm thick), 4 slices ginger, 60g carrot (sliced, 0.5 cm), 1 tsp canola oil

Seasonings:

1/4 tsp salt

Method:

1. Rinse and drain the ingredients before cooking; cut mustard cabbage into 4 cm segments, separating the leaves from stems (if stems are thick, cut diagonally).

2. Preheat an 11-inch skillet till water droplets "dance slowly"; add ginger, oil, cover; pan-fry for 1 min each, on both sides over low heat till fragrant, push to a corner; spread oil evenly, lay out stems, carrots, mushrooms evenly on 3 sections, cover; pan-fry for 1 min on medium low heat.

3. Turn to low heat, clean edge of pan with wet cloth, add the leaves loosely on top, sprinkle with salt, cover; pan-fry for 50 seconds on medium heat, till little steam appears; turn to low heat, give a quick stir till cooked thoroughly, and serve immediately.

Culinary Tips:

- Traditionally, the ingredients except the leaves are prepared by parboiling first, and then stir-fried, causing the veggies to lose its natural flavors; therefore salt, MSG or oyster sauce must be added.

- Our recipes emphasize the importance of maximum retention of the nutrients for every ingredient; so understanding their nature and cooking times is crucial in deciding the size of the cut, the pan heat, fire during cooking process and the sequence we place the ingredients into the pan to cook; in this case the leaves are placed into the pan after the other ingredients are partially cooked.

彩虹菜鍋

15分鐘 / 4人份 / 1人份55卡

15 mins / 4 servings / 55 cal per serving

材料：

綠花椰菜200克、白花椰菜200克、紅蘿蔔片4片
（0.8公分，共40克）、紅黃甜椒片各2片（約100
克）、玉米筍4根、秋葵4根、柳松菇1叢

調味料：

鹽少許

作法：

1. 綠色及白色花椰菜洗淨，取花朵狀（但不要太小
   朵）；玉米筍、秋葵、柳松菇分別洗淨，濾乾水
   分。

2. 取8吋平底鍋，先在底部鋪上一圈白花椰菜，再平
   均鋪上綠花椰菜，柳松菇放花菜中間，利用細縫
   處平均擺入紅甜椒及黃甜椒片、玉米筍、秋葵及
   紅蘿蔔片。

3. 烹煮前用水潤濕食材並濾乾，再撒入鹽，加過
   濾水1/2杯，轉中火煮約3～4分鐘至大量水氣冒
   出，用濕布擦鍋蓋，以食物刺針檢視熟度，熄火
   並馬上食用。

原味達人的烹調祕笈

- 傳統汆燙蔬菜時，習慣放入大量滾水燙熟；為了保
  持鮮綠與光澤度，還會添加小蘇打或油，燙熟撈起
  浸泡冰水降溫以增加脆度，如此易讓蔬菜的營養素
  流失二次。

- 一鍋料理多種蔬菜的原味烹調法，須先分辨食材熟
  度的屬性，再切成適當塊狀，慢熟的擺下面，快熟
  的如甜椒放上面，占滿鍋的空間再加蓋開火，讓上
  下對流的蒸汽循環均勻，瞬間使蔬菜同步煮熟，色
  澤更光鮮亮麗，並保留更多的葉綠素及脆度。

Ingredients:

200g broccoli, 200g cauliflower, 40g carrot
(sliced, 0.8 cm), 100g/2 slices each red and
yellow bell peppers, 4 baby corn, 4 gumbos
or lady fingers, 1 cluster brown swordbelt
mushrooms

Seasonings:

dash salt

Method:

1. Wash clean, drain, cut cauliflower and broccoli
   into flowerets (not too small). Separately
   wash and drain the baby corn, gumbos and
   mushrooms.

2. In an 8-inch skillet, lay out a circle of
   cauliflower on the bottom, then lay out
   evenly the broccoli on top and the cluster of
   mushroom in the center; within the flowerets,
   spread out red and yellow bell peppers, baby
   corn, gumbos and carrots on them.

3. Rinse with water, drain; sprinkle with salt,
   add 1/2 cup of filtered water; pan-boil for 3
   to 4 mins on medium heat, till much steam
   appears; wipe lid with wet cloth, test for
   doneness with the cooking pin, and serve
   immediately.

Culinary Tips:

- Traditionally, veggies are boiled and immersed;
  to preserve the fresh green and bright color
  of veggies, baking soda or oil is added; they
  are then immersed in cold water to keep
  them crisp, giving way to their nutrients to be
  lost twice.

- When cooking various kinds of veggies
  together through natural flavor cooking,
  there is a need to identify their nature and
  cooking times, and then cut in their relative
  proportions; if they are unable to cut to their
  proportion such as bell pepper, the ones that
  cook quicker are placed on top; filling the
  skillet almost full with the lid covered, enables
  the veggies to be cooked in a bath of vapor
  steam that is distributed evenly across the
  bottom and up the sides and instantly lock the
  natural qualities into the food, thus retaining its
  colors and crispness.

蒜香木須高麗菜

8分鐘 / 4人份 / 1人份56卡

8 mins / 4 servings / 56 cal per serving

材料：

高麗菜400克、木耳片120克、紅蘿蔔片100克（厚約0.8公分）、蒜頭3粒、紅辣椒片1根、芥花油1大匙

調味料：

鹽適量

作法：

1. 將高麗菜洗淨，濾乾水分，外葉剝成大塊，菜梗另外放。
2. 取12吋平底煎鍋，預熱至水珠慢跑，加入拍碎的蒜頭、芥花油，以小火爆香兩面各30秒至金黃色，取出；分邊放入菜梗、木耳片、紅蘿蔔片於鍋底，爆香好的蒜頭移至菜梗上，加蓋，以中小火炒1分鐘
3. 轉小火，開蓋用濕布擦鍋緣，稍微拌炒再入鹽、高麗菜葉及紅辣椒片，加蓋，轉中火燜炒約1分鐘至小水氣冒出來，轉小火，濕布擦鍋蓋再打開，並快速拌勻即成。

原味達人的烹調祕笈

傳統炒高麗菜就是用熱油翻炒，怕燒焦就加水燜煮至熟，要多油又多水，自然也要用較多的鹽及味精調味，反而吃不到高麗菜鮮甜的原味及營養。

蔬菜的原味烹調，首先須先了解食材的熟度屬性，將每個部位依先後順序烹煮的剛好熟時，可以減少調味料，保持清脆並吃出食材的真實原味。

Ingredients:

400g cabbage, 120g black wood-ear fungus, 100g carrot (sliced 0.8 cm), 3 garlic (crushed), 1 red chilly (sliced), 1 Tbsp canola oil

Seasonings:

dash salt

Method:

1. Wash and drain the cabbage before cooking, break the outer leaves into large pieces, setting aside the stem parts.
2. Preheat a 12-inch skillet till water droplets "dance slowly"; add garlic, oil, pan-fry for 1/2 min each, on both sides over low heat till golden yellow, remove; add cabbage stems, fungus, carrot and fragrant garlic on top, cover; pan-fry for 1 min on medium low heat.
3. Turn to low heat, uncover, clean edge of pan with wet cloth; stir slightly separately, add dash salt, cabbage leaves, red chilly, cover; pan-fry for 1 min on medium heat till little steam appears; turn to low heat, clean lid with wet cloth, give a quick stir, and serve immediately.

177

Culinary Tips:

· Traditionally, cabbage is prepared by stir-frying in much hot oil; to prevent the cabbage from getting burnt, much water, additional salt and MSG are added to enhance the flavor; thus the cabbage is neither crisp nor natural-flavored.

· By cooking the veggies with "waterless cooking", without additional water and cook sequentially various ingredients in the right order; basting the veggies in its natural moisture, enables us to reduce amount of seasoning intake and to savor the original flavor-filled taste of each ingredient, giving it a very pleasant crisp taste.

五福臨門

10分鐘 / 6人份 / 1人份63卡

Lucky Five

10 mins / 6 servings / 63 cal per serving

材料：

包心大白菜1棵（約600克）、薑片6片、綠花椰菜1棵（約250克）、紅蘿蔔片6片（約80克，厚0.8公分）、新鮮蘑菇6朵（約300克）、柳松菇1叢、芥花油1大匙

調味料：

味醂1大匙、鹽適量、香菇醬油2茶匙

作法：

1. 所有食材洗淨並濾乾，下鍋前用水潤濕；菜葉與白菜梗用刀切開成6公分長段；綠花椰菜切成6朵花狀。

2. 取12吋平底煎鍋，預熱至水珠慢跑，抹少許油，先把薑片放在鍋底，快速交錯疊放白菜梗，沿著鍋邊間隔放入綠花椰菜及蘑菇、紅蘿蔔片，中間放柳松菇。

3. 均勻撒入鹽及味醂，加蓋，轉中小火煮約1分鐘後，開鍋蓋，用濕抹布擦鍋緣；再交錯疊放白菜葉，加蓋，以中火煮約2～3分鐘至有大量水氣冒出來，熄火，以濕抹布擦鍋蓋，開蓋淋上醬油即成。

原味達人的烹調祕笈

- 傳統作法要把多種蔬菜先放入滾水中汆燙，再入油同鍋拌炒，不僅破壞其原味和原色又容易吃油，而造成需用較多的調味料提味或勾芡。

- 多種食材放在同鍋烹煮，只要依食材熟成屬性，切成合適大小，依序擺放，不需混炒，分次放入再同步煮至剛好熟透，便可以吃到每種食材的原汁原味。

Ingredients:

600g/1 Chinese white cabbage, 6 slices ginger, 250g/1 broccoli (6 flowerets), 80g carrot (6 slices, 0.8 cm thick), 300g/6 fresh mushrooms, 1 cluster brown swordbelt mushrooms, 1 Tbsp canola oil

Seasonings:

1 Tbsp mirin, dash salt, 2 tsp mushroom-flavored soy sauce

Method:

1. Rinse and drain the ingredients before cooking; separate the leaves and stems of white cabbage with a knife, cut into 6 cm segments.

2. Preheat a 12-inch skillet till water droplets "dance slowly"; coat skillet with oil, place ginger at the bottom, spread quickly the cabbage stems evenly in circles and staggered layers; along the rim, lay out the broccoli and button mushrooms alternately, in between place the carrots; place the cluster of swordbelt mushrooms in the center.

3. Sprinkle dash salt and mirin, cover; cook for 1 min on medium low heat, uncover, clean edge of pan with wet cloth, stagger the cabbage leaves in circles and layers on top of the stems, cover; cook for 2 to 3 mins on medium heat till much steam appears; turn off heat, clean lid with wet cloth, drizzle soy sauce, and serve immediately with the pan.

179

Culinary Tips:

· Traditional methods of cooking different vegetables together include parboiling first and then stir-frying with much oil, destroying both the flavor and color; additional seasonings and starch needs to be added to enhance the flavor.

· When cooking multiple veggies together, in order to retain the natural goodness of every ingredient, we must be familiar with the qualities of each kind of vegetable, be particular with the placement of veggies in the pan and to cook sequentially various veggies at separate times.

柴燒杏鮑菇 15分鐘 / 4人份 / 1人份84卡

Glazed King Oyster Mushrooms with Bonito Flakes

15 mins / 4 servings / 84 cal per serving

材料：

杏鮑菇3根（直徑5公分，共約600克）、烤酥
柴魚片10克（作法見P.168冰鎮柴魚菠菜）、
烤酥的白芝麻適量、芥花油2茶匙

調味料：

香菇醬油1/4杯、味醂1.5大匙

作法：

1. 杏鮑菇洗淨並擦乾。取11吋平底煎鍋，預熱
   至水珠慢跑，鍋底抹油並頭尾交錯放入整根
   的杏鮑菇，加蓋，以中火煎2分鐘（鐵板電
   烤鍋預熱至400°F，煎3分鐘），轉中小火
   （鐵板電烤鍋則預熱至350°F），煎其餘三
   面各約3分鐘。

2. 入調味料，加蓋煮開至小水氣冒出，轉小火
   （鐵板電烤鍋轉至275°F）續煮上下兩面各
   約1分鐘，讓兩面均勻上色，加入柴魚片，
   加蓋燜煮1分鐘，熄火。

3. 取出杏鮑菇，每根均等切成4長條，置入盤
   中，再鋪上柴魚片及烤酥的白芝麻，淋入鍋
   中剩餘的醬汁即成。

原味達人的烹調祕笈

- 杏鮑菇碩大肥厚的菇體較不好煎炒，大部分
  的人都會切片，放入滾水汆燙、裹粉油炸或
  是用油拌炒，如此一來易造成菇體縮小，失
  去彈性及鮮美的滋味，或使它的海綿組織吸
  取更多的油。

- 原味的烹調特色是盡量完整或大塊烹煮，小
  塊分食。將杏鮑菇整根清煎至快熟，只用少
  許醬汁及柴魚燒煮，可保留杏鮑菇更完整的
  原汁原味；食用時，切片或長條狀都有不同
  的口感，鮮嫩甜美又不油膩。

Ingredients:

600g/3 king oyster mushrooms (5 cm wide), 10g dried bonito flakes (pan-grilled, refer to page 168), dash white sesame seeds (pan grilled), 2 tsp canola oil

Seasonings:

1/4 cup mushroom-flavored soy sauce, 1.5 Tbsp mirin

Method:

1. Rinse, wash mushrooms and pat dry; preheat an 11-inch skillet till water droplets "dance slowly"; add oil, place mushrooms alternately, cover; pan-fry for 2 mins on medium heat (for electric griller, preheat to 400°F, grille for 3 mins); reduce to medium low heat to pan-fry for 3 mins each, for the other 3 sides (for griller, 3 mins each at 350°F).

2. Add seasonings, cover, pan-boil till little steam appears, turn to low heat (for griller, adjust to 275°F), simmer for 1 min each, glazing the mushrooms on both sides, sprinkle bonito flakes over the top, simmer for 1 min to let the aroma of bonito flakes permeate the mushrooms.

3. Cut into quarters horizontally, drizzle remaining sauce over the top with bonito flakes, garnish with sesame seeds and serve.

Culinary Tips:

· Due to its huge size, the king oyster mushroom does not cook easily; normally, it is being sliced, and prepared by parboiling, breading and deep-frying or stir-frying; so there is a tendency to absorb extra oil, cause food shrinkage and most important lose its nutritional value and natural juices.

· The characteristic feature of natural-flavor cooking is to cook the ingredient in whole, and is best to cut after cooking; pan-frying the mushroom in whole using minimal oil, allows the rich fragrance to emerge while retaining the natural juices; pan-fried till almost done and then glazed, eliminate the need for much sauce to cook the mushrooms; adding bonito flakes complement each other well and enables you to savor the blending of their succulent flavors and delicate smooth textures.

素煎雙菇 8分鐘 / 4人份 / 1人份92卡

Fresh Mushroom Surprise

8 mins / 4 servings / 92 cal per serving

材料：

新鮮香菇4朵（約400克）、大蘑菇4朵（約400克）、百里香適量、芥花油2茶匙

調味料：

鹽、黑胡椒粒各適量

作法：

1. 新鮮香菇、大蘑菇分別用過濾水沖淨，再擦乾水分，剪去少許蒂頭。

2. 取11吋平底煎鍋，預熱至水珠慢跑（鐵板電烤鍋預熱至400°F，烤約3分鐘），鍋底抹油，間隔放入香菇、大蘑菇（菇面朝下），撒鹽，以中火烤約2分半鐘至金黃色。

3. 轉小火，開蓋，以濕抹布擦鍋緣，再將香菇及大蘑菇翻面，撒鹽及黑胡椒，加蓋轉中小火烤約3分鐘至熟（鐵板電烤鍋預熱至350°F，烤約3分鐘），放入百里香燜煮約30秒即成。

百里香

原味達人的烹調祕笈

● 傳統烹調將菇類切半或切片，用熱油或奶油不停翻炒，香菇肉質容易出水而縮小，更會吸取鍋面油質而失去原味及彈性。

● 這道料理非常簡單又容易上手，只使用少許油煎整顆菇至表面呈金黃色，就可以封住菇表面的毛細孔及菇體內的原汁，煮到剛好的熟度，只要加入少許調味料，就可以吃到滑嫩彈Q的口感，很適合為了減肥而得犧牲美味的讀者。

Ingredients:

400g/4 fresh black mushrooms, 400g/ 4 white button mushrooms, dash thyme, 2 tsp canola oil

Seasonings:

dash salt, dash coarse black pepper

Method:

1. Rinse and wash clean the two kinds of mushrooms separately with filtered water, pat dry, trim off part of the stem ends.

2. Preheat an 11-inch skillet till water droplets "dance slowly" (for electric griller, preheat to 400°F, pan-grille for 3 mins); coat skillet with oil, place mushroom cap on pan bottom alternately with the button mushroom evenly, sprinkle with salt, cover; pan-fry for 2.5 mins on medium heat till golden yellow.

3. Turn to low heat, uncover, clean edge of pan with wet cloth, flip, sprinkle salt and pepper, cover; pan-fry for 3 mins on medium low heat till done (for griller, 3 mins at 350°F); turn to low heat, sprinkle thyme over the top, simmer for 1/2 min and serve immediately.

Culinary Tips:

· Traditionally, mushrooms are halved or sliced, stir-fried constantly in open cooking with much hot oil or butter; during constant stir-frying, it disables the pores to seal, thus absorbing more oil and releasing the natural juices into the oil causing the mushroom to shrink, the oil to splatter, and greatly destroying the flavor and texture.

· This is by far the simplest, easy-to-cook and prepare dish with only a dash of the natural seasoning salt and pepper; it is suitable for those who desire to remedy their poor diets and get rid of the extra pounds.

干炒四季豆

15分鐘 / 4人份 / 1人份121卡

ch Beans with Bean C

15 mins / 4 servings / 121 cal per serving

184

材料：

小豆干4塊（約200克）、四季豆300克、紅蘿蔔片80克（厚約0.5公分）、紅黃甜椒片各40克、蒜頭3粒、芥花油1大匙

調味料：

香菇醬油2茶匙、鹽及黑胡椒適量

作法：

1. 四季豆洗淨，去老絲並折對半；小豆干洗淨，切0.5公分厚片；大蒜拍碎。取12吋平底煎鍋，預熱至水珠慢跑，放入油、蒜頭，加蓋，以小火煎兩面各30秒，取出備用。

2. 一半鍋子內放小豆干，另一半放四季豆、紅蘿蔔片及蒜頭，在四季豆上撒鹽，加蓋，以中小火炒約2分鐘，至小水氣冒出，轉小火，開蓋，用濕抹布擦鍋緣，豆干翻面續煎，撒鹽在蔬菜上並鋪上甜椒。

3. 再加蓋，以中火炒約1分鐘至熟，轉小火，開蓋，小豆干燴入醬油，再撒上黑胡椒並翻炒均勻即成。

原味達人的烹調祕笈

- 傳統炒豆干通常會用較多的油及水翻炒，或先汆燙再用油快炒，容易破壞食物中的維生素及礦物質而縮水。

- 無油無水烹調是利用鍋具的恆溫產生均熱火候來炒菜，不需特別翻炒也美味可口。以寬的平底鍋面同步料理多種食材，用油省又可煎香豆干；四季豆因含水量不高，先以中小火炒至四季豆冒汗產生蒸汽，就不用另外加水烹調，也不容易有青澀味。

Ingredients:

200g/4 small dried tofu or bean curd (sliced,1/2 cm), 300g French beans, 80g carrot (half-sliced, 1/2 cm thick), 40g each red and yellow bell peppers (sliced, 2 cm wide), 3 garlic (crushed), 1 Tbsp canola oil

Seasonings:

2 tsp mushroom-flavored soy sauce, dash salt & coarse black pepper

Method:

1. Rinse, drain ingredients before cooking, de-vein French beans and manually rip in halves; preheat 12-inch skillet till water droplets "dance slowly"; add oil, garlic, cover; pan-fry for 1/2 min each, on both sides, remove.

2. Lay out bean curd on half of the skillet; the French beans, carrot and garlic on the other half, sprinkle salt on beans, cover; pan-fry for 2 mins on medium low heat till little steam appears; turn to low heat, uncover, clean edge of pan with wet cloth, flip, sprinkle salt on the veggies, add bell peppers on top.

3. Cover, pan-fry for 1 min on medium heat till done; turn to low heat, drizzle soy sauce over the bean curd, sprinkle pepper and give a quick stir.

Culinary Tips:

- Traditionally, bean curd and beans is prepared by quick frying in much oil and water; or parboiling and then stir fried, destroying the vitamins and minerals and resulting in food shrinkage.

- "Waterless and greaseless" cooking use different temperature settings or heat to give best results; by utilizing a wide skillet when cooking multiple ingredients together, the use of oil can be reduced and the bean curd can also be pan-fried till fragrant; also though the beans do not have a high moisture content, by cooking on medium low and then medium heat with the lid covered, the vapor seal method allows the veggies to be cooked in its own natural moisture without added water.

185

秋葵塔

5分鐘 / 4人份 / 1人份68卡

5 mins / 4 servings / 68 cal per serving

186

材料：

紅秋葵及綠秋葵各12個、黃色玉米1根

調味料：

鹽適量

作法：

1. 紅秋葵、綠秋葵洗淨並濾乾水分，底部不要切太多，以免流出黏液，下鍋前用水稍微潤濕；玉米洗淨，切成2公分圈狀。

2. 取8吋平底煎鍋，先將黃色玉米放在鍋中間，再順著鍋邊，嫩尖朝鍋上，平均交錯放入紅、綠秋葵，圍著玉米成圓塔狀。

3. 平均撒上鹽，再淋入過濾水120c.c.，加蓋，以中火煮約3分鐘至大量水氣冒出來，熄火，以濕抹布擦鍋蓋，開蓋即成。

原味達人的烹調祕笈

● 傳統料理秋葵常放入滾水中汆燙，撈起，再放入冰水漂涼，吃不到其特殊黏液。而且秋葵切斷烹煮，較難保留原味，還會滲出黏液，若煮過熟，口感太軟不好吃。

● 原味料理利用秋葵是一邊尖一邊寬的形狀，巧妙地使用玉米圈放在鍋中間，把秋葵尖角墊上，可避免秋葵尖角貼鍋煮過熟。以少許水烹煮，利用瞬間的蒸汽穿透，味道特別甜，還可保持黏液及脆感，原汁原味，不用加太多鹽，吃起來滑順又不澀。

Ingredients:

12 each red and green gumbos or lady fingers, 1 yellow corn

Seasonings:

dash salt

Method:

1. Cut off a little of the base of gumbo, retaining part of it to prevent loss of nutrients; rinse and drain the ingredients before cooking, cut corn into 2 cm rings.

2. Lay out the ring of corn in the center of an 8-inch skillet; along the edge of the bottom, lay out the red and green gumbos alternately, towered against the ring of corn.

3. Sprinkle salt evenly, add 120c.c. filtered water, cover; cook for 3 mins on medium heat till much steam appears; turn off heat, clean lid with wet cloth, uncover and serve immediately.

Culinary Tips:

· Traditionally, gumbos are parboiled and then immersed in iced water; thus losing its sticky juice within, giving it a soft texture and draining food of its natural goodness.

· The concept of the natural-flavor cooking is very simple; gumbo is shaped one end pointy and the other end wide; to prevent the pointy end of the gumbos from being overcooked, the gumbos are towered against the ring of corn, leaving the pointy ends above the corn; in this way the nutritional value can be retained, for a more flavor-filled gumbo with just a hint of salt and minimum water.

紅豆糯米甜湯

60分鐘 / 8人份 / 1人份381卡

Glutinous Red Be

60 mins / 8 servings / 381 cal per serving

材料：

紅豆400克、過濾水1000c.c.、白糯米1/2杯、熱水600c.c.、橙皮屑1/4杯、七葉蘭4片

調味料：

紅砂糖3/4杯、冰糖1/2杯、熱水1000c.c.

作法：

1. 紅豆洗淨，加入過濾水1000c.c.、七葉蘭（打成小結）一起放入4公升湯鍋，加蓋，以中火煮沸，轉小火燉煮約25分鐘。

2. 加入洗淨的白糯米及熱水600c.c.煮開，轉小火繼續燉煮約25分鐘至熟透。

3. 倒入砂糖及冰糖拌煮至融化，再加熱水1000c.c.煮沸，放入橙皮屑拌勻即成。

七葉蘭

Ingredients:

400g red beans, 1000c.c. filter water, 1/2 cup white glutinous rice, 600c.c. hot water, 1/4 cup orange zest, 4 pine or pandan leaves

Seasonings:

3/4 cup cane sugar, 1/2 cup rock sugar, 1000c.c. hot water

Method:

1. Wash the red beans; add 1000c.c. filtered water and pine leaves (tied into a knot) in a 4 quart pot; cover, bring to a boil on medium heat, turn to low heat and simmer for 25 mins.

2. Add washed glutinous rice and 600c.c. hot water; bring to a boil again, turn to low heat and simmer for another 25 mins till cooked.

3. Add sugar, stir till sugar is thoroughly melted, add 1000c.c. of hot water and bring to a boil; add in orange zest and mix well.

原味達人的烹調祕笈

○ 傳統煮紅豆會先浸泡4～6小時，使紅豆體積膨脹，但入電鍋煮易破裂，湯汁較混濁；若用壓力鍋煮，雖可省時，但紅豆在湯汁中滾來滾去，易失去香氣及紅豆的原味，口感也較鬆軟不綿密；而選用燜燒鍋則少了細火慢燉的香氣。

○ 要保持豆子的原味與香氣，要先用少許水煮至半透再入熱水及快熟的白糯米繼續煮至紅豆熟透，最後加入糖及熱水滾沸即可吃到顆粒紅潤，口感香甜綿密的紅豆。

○ 橙皮有特殊香氣，屬鹼性，可平衡豆類酸性，吃起來不致造成胃酸或太膩。

Culinary Tips:

· The traditional method of preparing red beans is to immerse in water for 4~6 hours so that the beans will expand; however, once cooked, the beans crack easily, clouding the soup. Using a pressure cooker may save time, but the beans trundle in the water throughout the cooking process, losing their natural flavor and also become too soft and loose; using a smolder cooker will not emerge the scent as there is no simmer heat.

· To preserve the natural flavor and aroma of the red beans, they are cooked with just enough water to simmer; the glutinous rice which cook faster is added subsequently with also just enough hot water for the rice to be cooked; the balance of the hot water and sugar is added at the end.

· The orange zest adds a special fragrance and also serves to balance the acidity of this dessert, so as to prevent gastric acid.

189

紅酒燉牛腩

40分鐘 / 8人份 / 1人份263卡

40 mins / 8 servings / 263 cal per serving

材料：

加拿大牛腩600克、大番茄3顆（約400克）、洋蔥1顆（約300克）、紅蘿蔔1條（約120克）、紅馬鈴薯4顆（約400克）、紫色高麗菜半顆（約400克）、西洋芹2根（約200克）、蔥白3段、薑6片、八角1粒、月桂葉2片

調味料：

香菇醬油1/2杯、紅酒1/2杯

作法：

1. 牛腩先剪除多餘脂肪，以檸檬皮和粗鹽搓洗濾乾；番茄、洋蔥及紅蘿蔔洗淨，對切；紅馬鈴薯洗淨不去皮；紫色高麗菜切成四等分；西洋芹洗淨，切成2段。

2. 取3公升平底湯鍋，預熱至水珠快跑，放入牛腩，以中火乾爆兩面至小水氣冒出呈金黃色，開蓋轉小火，用濕布擦鍋緣再加入薑片、蔥段及八角以中小火爆香1分鐘。

3. 轉小火，倒入紅酒、醬油、月桂葉、番茄、洋蔥、紅蘿蔔及馬鈴薯，加蓋，以中火煮沸，轉小火燉煮約25分鐘，放入紫色高麗菜、西洋芹，轉中火煮沸，以小火續煮約5分鐘即成。

原味達人的烹調祕笈

無油烹調牛腩是整條烹煮到剛好熟後，再取出切塊分食，可保留原汁原味，吃起來香嫩又Q又不腥。配合牛腩，所有蔬菜依其屬性切成大塊烹煮，小塊分食，可保留較多原味及養分，也不需太多調味料提味。

無油、無水、無鹽、無糖料理的牛腩條，烹煮時間恰當就不會軟爛失味，每樣食材都維持原味，又有混煮後的特殊風味，食用時再將牛腩切小塊，可品嚐到真實的原汁原味。

Ingredients:

600g Canadian beef rib finger, 400g/3 large tomatoes (halved), 300g/1 onion (halved), 120g/1 carrot (halved), 400g/4 red potatoes (unpeeled), 400g red cabbage (quartered), 200g/2 stalk celery (segmented), 6 slices ginger, 3 stalk scallion (segmented), 1 star anise, 2 bay leaves

Seasonings:

1/2 cup mushroom-flavored soy sauce, 1/2 cup red wine

Method:

1. Trim off excess fat from beef; wash beef with lemon skin and sea salt; rinse, drain and pat dry.

2. Preheat a 3-quart pot till water droplets "dance quickly"; pan-broil beef three sides on medium heat till little steam appears and golden brown; turn to low heat, clean edge of pan with wet cloth, add ginger, scallion, star anise; pan-fry for 1 min on medium low heat till fragrant.

3. Turn to low heat, add wine and soy sauce, top beef with bay leaves, tomatoes, onion, carrot and potatoes, cover; bring to a boil on medium heat, reduce to low heat, simmer for 25 mins; add cabbage and celery, bring to a boil on medium heat, simmer for another 5 mins on low heat and serve.

Culinary Tips:

· The main concept of this recipe is to pan-broil the beef to force out excessive fat; then the spices are fried with no additional oil; the vegetables are simmered without parboiling first, and only a small amount of seasoning is necessary, preserving all the natural nutrients.

· Cooked using the greaseless and waterless method, the beef in this sugar-free and salt-free stew tastes tender with a bite to it; the beef is not cut beforehand, but rather after it is cooked and the potatoes are cooked with skin; cooked along with larger pieces of vegetables and with the perfect timing, each ingredient maintains its natural and original flavor. The combined flavor of all ingredients is unique in its own, too. Wait until just before eating to cut into smaller pieces, and you will be able to taste the authentic flavor of each ingredient. It can be served as a wonderful all-in-one meal with multi-grain bread.

191

蘿蔔封肉

35分鐘 / 6人份 / 1人份406卡

Braised Pork with Carrot and White Radish

35 mins / 6 servings / 406 cal per serving

192

材料：

梅花肉600克、番茄1/2顆（約150克）、洋蔥1/2顆（約150克）、白蘿蔔720克、紅蘿蔔360克、蒜頭2粒、蔥段2根、薑片4片、香菜少許

調味料：

米酒1大匙、香菇醬油3/4杯

作法：

1. 全部食材洗淨，梅花肉用檸檬皮及粗鹽搓洗，濾乾；切成每塊重約100克；紅蘿蔔去皮，切成每塊重約60克；白蘿蔔去皮，切成每塊重約120 克；如果白蘿蔔的直徑超過8公分時，就要切成半圓的塊狀，使重量仍維持120克。

2. 取3公升平底深鍋，預熱至水珠快跑，放入梅花肉貼鍋以中火乾爆三面至小水氣冒出呈金黃色，轉小火翻面，放入薑片、蒜頭、蔥段以中小火爆香至金黃色。

3. 趁熱倒入米酒、醬油於肉上，依序放入番茄、洋蔥、白蘿蔔、紅蘿蔔，加蓋轉中火煮沸後，再轉小火燜煮約25分鐘，開蓋以食物刺針檢測熟度，熄火點綴香菜即成。

原味達人的烹調祕笈

- 乾爆梅花肉，可去腥並逼出多餘油脂，封住表面毛細孔，利用逼出的油脂爆香辛香料，在密閉的空間，以最少的醬汁，讓食材烹煮時產生的蒸汽可以互相對流，並充分滲透至中間食材，使蘿蔔外形完整，內部熟透，口感清甜卻帶有少許的肉香味，而梅花肉口感香Q又嫩。

- 紅白蘿蔔依熟度屬性，切合適的塊狀；白蘿蔔擺在肉上，而紅蘿蔔放最上層，不加水，不加糖，紅燒煮至剛好透，白蘿蔔吃起來清甜多汁；紅蘿蔔口感極似地瓜。

Ingredients:

600g pork picnic shoulder, 150g/ 1/2 tomato, 200g/ 1/2 onion, 720g white radish, 360g carrot, 4 slices ginger, 2 cloves garlic (peeled), 2 scallion stalk (segmented), dash Chinese parsley

Seasonings:

1 Tbsp rice wine, 3/4 cup mushroom-flavored soy sauce

Method:

1. Wash pork with lemon skin and sea salt, rinse, drain and pat dry; cut pork into 6 pieces, 100g each; cut carrots into 60g rounds and radish into 120g rounds; if the width of the radish is more than 8 cm thick, then it should be cut into semi-cylinder-shaped block, carvying the same weight 120g.

2. Preheat a 3-quart pot till water droplets "dance quickly"; pan-broil three sides on medium heat till little steam appears and golden brown; turn to low heat and flip on every side, cleaning edge of pot with wet cloth, add ginger, garlic and scallion; pan-fry on medium low heat till fragrant.

3. Drizzle rice wine and soy sauce over the pork, add in tomato, onion, radish, and carrot, cover; bring to a boil on medium heat, then simmer on low heat for 25 mins; test for doneness, garnish with Chinese parsley.

Culinary Tips:

· Cooked using the greaseless and waterless method and filling the pot with ingredients to the brim, prevent oil, water and air to rob food of its nutrition; this process cooks food in its own vapor steam, keeping the aroma in the pot. Pan-broiling the pork eliminates the meaty scent, seal up the pores thus expanding the meat and force out excessive fat to fry the spices so as to preserve the traditional flavor of this dish.

· Cutting each ingredient proportionately according to their nature, and cooked with the perfect timing allows all ingredients to be naturally sweet and delicious without adding sugar, salt and much sauce; also the pork taste tender and the radish and carrot very juicy; laying the carrots on the very top retains its sweet potato-like taste.

彩椒蘆筍燴鮮貝

15分鐘 / 4人份 / 1人份216卡

Pan-Seared Scallops with Colorful Peppers and Asparagus Sauté

15 mins / 4 servings / 216 cal per serving

材料：

加拿大野生干貝8顆（約200克）、大蘆筍6根（約350克）、紅黃甜椒各約100克、新鮮香菇3朵（約150克）、蒜頭2粒（15克，拍碎）、芥花油3茶匙

醃料：

蔥薑蒜酒汁2茶匙

調味料：

鹽及黑胡椒適量

作法：

1. 所有食材洗淨；紅黃甜椒切成2公分寬的斜片；大蘆筍切成4公分的長段；香菇切成2公分的厚片；干貝用蔥薑蒜酒汁醃30分鐘後擦乾；所有食材下鍋前先用水潤濕。

2. 取11吋平底煎鍋，預熱至水珠慢跑，加入蒜頭及2茶匙的油，以小火爆香兩面成金黃色，取出，再依序加入蘆筍、香菇及爆香好的蒜頭，加蓋，小火約1分鐘，輕輕搖鍋數下，轉中小火約1分鐘。開蓋，入甜椒及調味料拌炒，加蓋以中火約1～2分鐘至熟。

3. 利用炒蘆筍的時間，另取8吋平底煎鍋，預熱至水珠快跑，加入1茶匙的油抹勻，貼鍋放入干貝，撒少許鹽及黑胡椒，以中火煎約1分30秒至金黃色，轉小火，翻面撒鹽及黑胡椒，加蓋轉中小火續煎約1分30秒，取出放在蘆筍上即成。

原味達人的烹調祕笈

- 大蘆筍的頭部較窄，底部較厚，須依熟度的屬性切段，讓每塊蘆筍剛好熟。新鮮香菇的海綿組織易吸油，不放鍋底煮，改放蘆筍上面，利用烹煮時產生的蒸汽燻熟，沒有菇的腥味，保留其原汁，吃起來特別的滑Q。

- 干貝的屬性與蔬菜不同，因此同時使用兩個鍋料理，利用蘆筍加蓋煮時，另取小煎鍋，把干貝煎到金黃色又剛好熟。

Ingredients:

200g/8 Canadian wild scallops, 350g/6 large asparagus (4 cm segments), 100g each red and yellow bell peppers (sliced 2 cm), 150g/3 fresh black mushrooms (sliced 2 cm), 15g/2 cloves garlic (crushed), 3 tsp canola oil

Marinade:

2 tsp mixed juice of scallion, ginger, garlic and wine

Seasonings:

dash salt, dash coarse black pepper

Method:

1. Marinade scallops for 30 mins, pat dry; rinse ingredients before cooking.

2. Preheat an 11-inch skillet till water droplets "dance slowly"; add garlic and 2 tsp oil, pan-fry both sides on low heat till golden yellow, remove; add asparagus, mushroom and garlic on top, cover; pan-fry for 1 min on low heat and 1 min on medium low heat, shaking the pan to and fro occasionally; uncover, add bell peppers and seasonings, stir, cover; pan-fry for 1~2 mins on medium heat till done.

3. While frying asparagus, preheat an 8-inch skillet till water droplets "dance quickly"; grease pan with 1 tsp oil, place scallops evenly, sprinkle salt, cover; pan-fry for 1.5 mins on medium heat till golden yellow; turn to low heat, flip, sprinkle black pepper, cover; pan-fry for 1.5 mins on medium low heat till flesh is opaque; top asparagus with scallops.

195

Culinary Tips:

· The tip part of choice asparagus is thinner while the end is thicker; therefore, asparagus should be cut proportionately so that each piece can be cooked evenly with the right timing; fresh mushrooms, when cut absorb grease easily; place them on the asparagus, and allow the food to cook completely in its own natural moisture; retaining their natural juices; also the mushrooms taste smoother.

· As the nature of scallop is unlike asparagus, using a separate skillet to pan sear the scallop with a little oil till golden yellow, lock in the natural moisture and emerge its fragrance giving a tender, delicate and naturally juicy flavor.

原味鮮貝盅

10分鐘 / 4人份 / 1人份230卡

10 mins / 4 servings / 230 cal per serving

材料：

大蛤蜊8粒（約1000克）、洋蔥片1/2顆（約150克）、西洋芹1根（約100克）、紅黃甜椒片80克

作法：

1. 蛤蜊先以棕櫚刷清洗外殼，放入不鏽鋼盆中，倒入剛好淹過蛤蜊的冷水；加入粗鹽1大匙及不鏽鋼筷或小湯匙，放置於陰暗處數小時，使其充分吐沙。
2. 將洋蔥、西洋芹菜及甜椒切成2公分寬片、洋蔥保留三層的厚度。
3. 取8吋平底煎鍋，放入已充分吐沙的蛤蜊，再利用空隙處，放入洋蔥、西洋芹及甜椒，加蓋，以中火煮約4～5分鐘至大量水氣冒出至剛好熟，熄火即成。

原味達人的烹調祕笈

● 蛤蜊拌炒容易縮水，有些作法會先入滾水汆燙再炒，或直接放入炒鍋中拌炒，再將蛤蜊釋放出來的汁液，加入太白粉勾芡，使芡汁附著蛤蜊肉，很難吃到蛤蜊的鮮甜原味。

● 大蛤蜊以無油無水乾烤法，可直接逼掉海腥味，再加入蔬菜香味嗆入海鮮，不會只有單純的海鮮味。蛤蜊烤到剛好熟，肉質飽滿不縮水，不加任何調味料，味道就鮮美無比。

● 獨特的無油無水料理烹調法，可以保留營養到最高點，去油去脂又可品嚐到食物的原汁原味。

Ingredients:

1000g/8 large ocean clams, 150g/ 1/2 onion, 100g/1 stalk celery, 80g red & yellow bell peppers

Method:

1. Wash clams, immersed in chilled filtered water with 1 Tbsp of sea salt for a few hours in a dark corner, to release the sand within thoroughly and drain.
2. Sliced onion, celery and bell peppers 2 cm thick; for sliced onion, leave 3 layers intact.
3. Place the clams in an 8-inch skillet, and top with onion, celery and bell peppers in between the clams; cover, pan-broil for 4~5 mins on medium heat till much steam appears, test for doneness; serve immediately.

Culinary Tips:

· Traditionally clams are sometimes parboiled first, and then stir-fried; or stir-fried directly and add starch to thicken the juice released from the clams; both methods cause the clams to shrink and lose its natural flavor and nutrients.

· By pan-broiling the clams, the ocean sea water scent is eliminated, while the veggies that are cooked with them will flavor the clams during the process; cooked with the perfect timing, the clams will not shrink, and no additional flavoring needs to be added.

· The effective cooking process of the unique "greaseless and waterless" cooking retains the abundant nutrients, aroma and natural flavor-filled taste of food while reducing the calories and fat; therefore keep healthy cooking as the focal point.

鮮烤黃金蟹 15分鐘 / 2人份 / 1人份502卡

15 mins / 2 servings / 502 cal per serving

198

材料：

加拿大深海黃金蟹1隻（約800克）、老薑4
片、香油1茶匙

醃料：

蔥薑蒜酒汁1大匙

調味料：

鹽及黑胡椒滴量、檸檬辣椒醬適量

作法：

1. 黃金蟹用棕櫚刷洗淨後，去除腹底外殼及泥
   腸，再將外殼與身體分開並取出內臟及氣
   室，保留蟹膏；最後用過濾水沖洗並濾乾。

2. 取11吋平底煎鍋，預熱至水珠慢跑，放入香
   油1茶匙及薑片，以小火爆香兩面各30秒，
   先加入蟹身（腹部朝下）及蟹鉗，薑片放
   回蟹身上，以中火烤約3分鐘（或電烤鍋以
   400°F烤5分鐘）。

3. 轉中小火，開蓋，翻面撒鹽及黑胡椒，加蓋
   以中小火續烤約1分30秒，放入蟹殼於鍋底
   （內朝上）再烤約1分30秒至熟（電烤鍋以
   400°F烤3分鐘）；食用前以專用壓碎器夾蟹
   腳取肉，原味或沾食檸檬辣椒醬皆美味。

原味達人的烹調祕笈

- 此道使用加拿大螃蟹，氣室較乾淨（也可整隻
  帶殼煮），且肉質細嫩、味道鮮美、腳肉較紮
  實。保存方法同龍蝦。

- 傳統料理螃蟹，若要味道香、肉質Q，通常採
  油炸，肉質容易流失甜分又縮水；若選擇清
  蒸，但味道較不香；若是茄汁螃蟹則會使用很
  多番茄醬，有時還會勾芡調味，再淋在炸好或
  蒸好的螃蟹上面，卻吃不到螃蟹的原味。

- 螃蟹本身屬寒性食物，使用少許香油及薑片爆
  香，可驅寒補身。帶殼烤螃蟹，其蟹膏較快
  熟，得分開煮；以鹽及黑胡椒清烤，可保留蟹
  膏的香味，蟹肉鮮嫩多汁，不需沾醬也很美
  味。

Ingredients:

800g/1 Canadian golden crab, 4 slices aged ginger, 1
tsp sesame oil

Marinade:

1Tbsp mixed juice of scallion, ginger, garlic and wine

Seasonings:

dash salt, dash black pepper, dash lemon chilly sauce

Method:

1. Clean crab with palm brush, twist and pull tail flap
   off, pry up the top shell and remove greyish bag,
   pull out mandibles, remove the gills from body,
   twist off claws, rinse and drain; marinade for 1/2
   hour, pat dry.

2. Preheat an 11-inch skillet till water droplets "dance
   slowly"; add sesame oil and ginger, fry both sides
   for 1/2 min on low heat, remove; place the crab
   belly side down, claws, and ginger on top, pan-grille
   for 3 mins on medium heat.

3. Turn to low heat, uncover, flip, sprinkle salt and
   pepper, cover; grille for 1.5 mins on medium low
   heat; place the crab shell, up-side down on the pan
   and grille for 1.5 mins till done (for electric griller,
   grille at 400°F for 3 mins throughout); crack the
   claws with a nutcracker; the lemon chilly sauce can
   be served as a dip for this dish.

Culinary Tips:

- Canadian crab, whose gills is cleaner, can be cooked
  in whole, retaining its abundant nutrients and flavor;
  preserve it the same way as the lobster.

- Traditionally, crab is usually deep-fried, causing the
  meat to shrink and lose its natural goodness; as a
  healthier choice, crab is steamed, but it is not as
  flavorful; if the crab is prepared with a tomato-
  based recipe, much ketchup is required and starch
  is added to thicken the sauce and then poured over
  the crab; thus the natural taste of the crab is barely
  tasted.

- Pan-grilled with a small amount of sesame oil and
  ginger, it warms our body, emerge its fragrance and
  the distinctive flavor of the crab; using only salt and
  pepper will preserve the natural flavor and is enough
  to give it a succulent and delicate aroma, even
  without the dip.

新加坡海南雞 40分鐘 / 10人份 / 1人份488卡

Singaporean-Style Hainanese Chicken

40 mins / 10 servings / 488 cal per serving

200

材料：

母玉米雞1隻（約2000克）、蔥白段3根、老薑120克（拍碎）、蒜頭1球（約50克）、小黃瓜片100克、番茄片100克

調味料：

米酒1大匙、香油1茶匙、鹽3/4茶匙

配飾及沾醬：

香菜段少許、青蔥絲及辣椒絲適量、檸檬辣椒醬適量

作法：

1. 用檸檬皮及粗鹽逆向搓洗雞身，以熱水氽燙雞腹再塞入青蔥；取4公升湯鍋，加入過濾水600c.c.煮沸，加進蔥白段、去外膜後的蒜頭、老薑及玉米雞（雞胸朝上，頭及脖子反折向下，採仰泳姿勢），以中火煮開。

2. 轉中小火煮約16分鐘，熄火，續燜約16分鐘；取出全雞，淋上米酒，然後抹少許香油，再均勻塗抹鹽於雞皮上。

3. 待雞冷卻後，去骨切片，依序將四肢自關節處切除，接著取胸肉和腿肉切成1公分厚片，雞脖子切小塊，取出的雞骨架等可熬煮成雞湯底。食用時，配上小黃瓜片、番茄片，點綴香菜段、蔥絲及辣椒絲，也可用檸檬辣椒醬做為沾醬，更美味。

原味達人的烹調祕笈

● 傳統煮雞常常整隻入鍋，胸肉朝下，背骨朝上，滿水熬煮，雞胸肉會較澀，甜分也容易流失。

● 取適當鍋具，在剛好的空間，利用少水半蒸煮的方式，雞胸朝上雖沒碰到水，但蒸汽仍可穿透至熟，雞胸肉的味道甜而不澀。當整隻雞煮到剛好熟再抹上調味料，不需味精或雞粉，皮Q肉嫩又飽含雞汁，即使雞肉冷掉，口感仍滑Q鮮甜；煮雞的雞湯則像濃縮雞精般香濃。

Ingredients:

2000g/1 corn-fed female chicken, 3 stalks scallion (segmented), 120g aged ginger (crushed), 100g/1 garlic bulb (peeled), 100g cucumber (sliced), 100g tomato (sliced)

Seasonings:

1 Tbsp rice wine, 1 tsp sesame oil, 3/4 tsp salt

Garnish & Dip:

dash Chinese parsley (chopped), scallion (shredded), red chilly (shredded), dash lemon chilly sauce

Method:

1. Clean chicken thoroughly with lemon skin and sea salt, scald stomach with boiling water, place scallion green in stomach; bring to a boil with 600 c.c. filtered water in a 4-quart pot; add scallion stalk, garlic and ginger; place chicken breast-side up and bend neck and head down; bring to a boil on medium heat.

2. Cook chicken for 16 mins on medium low heat; turn off heat, set aside covered for 16 mins, test for doneness; remove and drain chicken, drizzle rice wine, sesame oil and rub evenly with salt on chicken skin.

3. Cool down chicken; cut neck, wings, drumstick and legs off at the joints; debone breast and drumstick and cut into 1 cm slices, chopped neck into pieces, the backbone can be boiled and become a base for soup; garnish and serve with lemon chilly sauce.

201

Culinary Tips:

· Traditionally, the whole chicken is cooked immersed in water, breast-side down; losing its natural moisture and sweetness; it is usually served with much sauce or MSG for a more flavorful and juicy taste.

· Selecting a right size pot with perfect timing for the chicken with minimum water, and placing the breast-side up, retains the natural moisture to tenderize the whole chicken including the breast; it also gives the chicken skin a chewy texture.

乾煎鮭魚香椿醬炒飯

20分鐘 / 4人份 / 1人份371卡

Fresh Cedar Garlic Rice with Pan-Broiled Salmon

20 mins / 4 servings /
371 cal per serving

材料：

加拿大鮭魚300克（厚2.5公分）、野米飯3杯、洋蔥丁150克、紅蘿蔔丁80克、蘆筍丁120克、紅黃甜椒丁各50克、烤香松子50克、芥花油2茶匙

醃料：

蔥薑蒜酒汁2茶匙

調味料：

鹽及黑胡椒適量、香椿醬1/4杯

作法：

1. 鮭魚用檸檬皮及粗鹽洗淨並濾乾，加入蔥薑蒜酒汁醃約30分鐘，入鍋前擦乾。取8吋平底煎鍋，預熱至水珠快跑，將鮭魚皮朝下貼鍋輕壓至逼出少許油，撒入鹽，加蓋以中火煎約1分鐘，轉中小火，續煎約2分鐘，翻面撒入少許鹽及黑胡椒，續煎約2分鐘，起鍋，用叉子剝成小片狀。

2. 煎魚時，另取11吋平底煎鍋，預熱至水珠慢跑，倒入1茶匙的油抹勻，鍋內分邊放入洋蔥丁、紅蘿蔔丁及蘆筍丁，轉中小火，加蓋，炒約1分鐘。

3. 開蓋快速拌炒，續入紅黃甜椒丁，撒鹽再鋪上野米飯，放入香椿醬，加蓋，轉中火約1分鐘後，開蓋拌勻，起鍋撒入烤香的松子及鮭魚片即成。

原味達人的烹調祕笈

- 要煮出好吃的原味料理，首選健康食材。此道選用加拿大有機野米，其蛋白質含量比一般白米多50%，還有多種維生素及礦物質。
- 野米煮法：野米1.5杯洗淨入3公升鍋，再加入過濾水4.5杯，浸泡約50分鐘，先以中火煮沸，轉小火燉煮約40分鐘。
- 鮭魚皮朝下乾煎可逼出多餘油脂並保留魚肉原汁。利用蔬菜本身的蒸汽將飯燻鬆，開蓋拌炒，可保持飯香又減少用油量，不需一直翻炒。飯炒香後，再把煎好的鮭魚片鋪在米飯上，可享受到較多的鮭魚原味及營養素。

Ingredients:

300g Canadian salmon (2.5 cm thick), 3 cups Canadian wild rice (cooked), 150g onion (diced), 80g carrots (diced), 120g asparagus (diced), 50g each red and yellow bell pepper (diced), 50g pine nuts (pan-grilled), 2 tsp canola oil

Marinate:

2 tsp mixed juice of scallion, ginger, garlic and wine

Seasonings:

dash salt, dash black pepper, 1/4 cup cedar sauce

Method:

1. Clean salmon with lemon skin and sea salt, drain, marinate for 30 mins, pat dry; preheat an 8-inch skillet till water droplets "dance quickly"; place salmon skin-side down in pan, pressing down till traces of fat is being forced out, sprinkle salt, cover; pan-broil for 1 min on medium heat, reduce to medium low heat for 2 mins; flip, sprinkle salt and black pepper, pan-broil for 2 mins till desired doneness; using a fork, break it into slices on fried rice.

2. Preheat an 11-inch skillet till water droplets "dance slowly"; add oil, place onion, carrot and asparagus separately, cover; pan-fry for 1 min on medium low heat.

3. Give a quick stir, add peppers, sprinkle salt, top with cooked wild rice and cedar sauce, cover; pan-fry for 1 min on medium heat, stir well; garnish with pine nuts and salmon.

Culinary Tips:

- This recipe uses organic wild rice which includes many vitamins and minerals, is more nutritious; the protein content is 50% higher than white rice. To cook 1.5 cups wild rice: wash, then soaked in 4.5 cups of filtered water in a 3-quart pot for 50 minutes, bring to a boil on medium heat and simmer for 40 mins. on low heat till water is completely absorbed into rice.
- Pan-broil salmon skin-side down, forces out excessive fat and eliminate the fishy scent, at the same time preserves its natural moisture; top half cooked ingredients with wild rice allows the vapor steam to baste food, retains natural flavors and keeping the aroma in the pan.

原味什錦炒米粉

30分鐘 / 8人份 / 1人份329卡

Original-Flavored Assorted Vegetables Fried Rice

30 mins / 8 servings / 329 cal per serving

材料：

香菇絲30克、乾蝦米30克、蔥白段4支、梅花肉絲200克（厚0.5公分）、米粉1包（350克）、洋蔥絲150克（直切寬1公分）、紅蘿蔔絲120克（0.5公分）、木耳絲100克（0.5公分）、高麗菜絲600克（葉2公分寬，菜梗1公分寬）、豆芽菜150克、芹菜末80克、紅蔥酥2大匙、芥花油2大匙

醃料：

香菇醬油1茶匙、香油1/2茶匙、蠔油1茶匙、白胡椒粉1/2茶匙、蔥薑蒜酒汁1茶匙

調味料：

香菇醬油80c.c.、黑胡椒適量

作法：

1. 米粉泡冷水20分鐘，濾乾水分後撥散；肉絲加入醃料拌勻10分鐘；其他材料在卜鍋前用水潤濕。

2. 取12吋平底煎鍋，預熱至水珠慢跑，放油1茶匙、香菇、蝦米、蔥白，加蓋，以中小火爆香兩面各1分鐘，取出；再入油1茶匙、肉絲並攤平，加蓋以中火爆炒1分鐘，翻面鋪上香菇、蝦米、蔥白，加蓋約半分鐘，快速拌炒轉小火，取出。

3. 再入油1大匙，加洋蔥、木耳一起拌炒，再入紅蘿蔔及高麗菜梗拌勻，依序鋪上高麗菜絲、米粉及醬油，加蓋，轉中火炒至大量水汽冒出，開蓋，再加黑胡椒及豆芽菜，快迅拌炒，入作法2材料一起拌勻，撒芹菜末及紅蔥酥即成。

原味達人的烹調祕笈

傳統炒米粉先將米粉浸泡熱水，使用較多的油水醬汁翻炒米粉及配料，米粉容易吸油水，不能完全蒸透又油膩，吃多會胃脹氣、消化不良。

用無水烹調，只須少量油爆香料，再藉由炒菜產生的蒸氣，把米粉燜至完全熟透，較容易融合菜料，使米粉香Q甘甜不油膩。

Ingredients:

30g dried mushrooms (shredded), 30g dried sea shrimps, 4 scallion stalk (segmented), 200g pork picnic shoulder (0.5 cm shreds), 350g/1 packet rice- flour noodles, 150g onion (shredded vertically, 1 cm), 120g carrot (0.5 cm shreds), 100g black fungus (0.5 cm shreds), 600g cabbage (2 cm shreds, 1 cm for stem), 150g bean sprouts, 80g Chinese celery (minced), 2 Tbsp fried shallots, 2 Tbsp canola oil

Marinade:

1 Tbsp mushroom-flavored soy sauce, 1/2 tsp sesame oil, 1 tsp oyster sauce, 1/2 tsp white pepper, 1 tsp mixed juice of scallion, ginger, garlic and wine

Seasonings:

80 c.c. mushroom-flavored soy sauce, dash black pepper

Method:

1. Soak rice noodles in cold filtered water for 20 mins, drain and loosen noodles; marinate pork for 10 minutes; rinse ingredients before cooking.

2. Preheat a 12-inch skillet till water droplets "dance slowly"; add 1 tsp of oil, place mushrooms, shrimps and scallion, cover; pan-fry each side for 1 min on medium low heat, remove; add 1 tsp oil, spread pork evenly, cover; pan-fry for 1 min on medium heat; flip, top with mushrooms and shrimps, cover; pan-fry for 1/2 min on medium heat, give a quick stir, turn to low heat, remove.

3. Add 1 Tbsp oil, stir in onion, black fungus, then carrot and cabbage stems, top with cabbage leaves, then rice noodles, drizzle soy sauce, cover; cook on medium heat till much steam appears; uncover, add black pepper, bean sprouts, mixed well, stir in ingredients in step 2; garnish with celery and shallots.

Culinary Tips:

· Traditionally, rice noodles are soaked in hot water or parboiled; also much oil and water is added to constantly stir-fry the noodles and ingredients, causing the noodles to be soft and greasy.

· Cooked using the waterless method and minimum oil, the vapor steam cook the noodles in its own natural moisture on top of the veggies, keeping the aroma in the pan, and giving the noodles a nice refreshing taste with an elastic texture that is not too soft.

南瓜船

15分鐘 / 4人份 / 1人份282卡

15 mins / 4 servings / 282 cal per serving

206

材料：

黃金小南瓜600克、梅花肉絲150克（0.5公分）、乾蝦米40克、香菇絲20克、蒜末15克、芥花油2茶匙、青蔥末及紅蔥酥各適量

醃料：

香菇醬油2茶匙、蠔油1茶匙、香油1/2茶匙、白胡椒粉1/2茶匙、蔥薑蒜酒汁1茶匙

作法：

1. 南瓜外皮洗淨，直切成7公分寬的船型；肉絲加醃料拌勻10分鐘。取2公升鍋，放入小南瓜（皮朝下）及水100c.c.，以中火煮沸至大量水氣出，轉小火續煮約6～8分鐘至熟。

2. 趁南瓜烹煮時，另取8吋小煎鍋，預熱至水珠慢跑，加入蒜末及芥花油1茶匙，以小火爆香至金黃色，取出，加入蝦米、香菇及蒜酥，加蓋，以中小火爆香兩面各約1分鐘，轉小火並取出。

3. 入油1茶匙，加入肉絲並鋪平，加蓋，以中火爆炒1分鐘，開蓋翻面鋪上作法2材料，加蓋以中火炒約30秒，開蓋快速拌炒均勻，起鍋鋪在南瓜上撒適量的青蔥及紅蔥酥即成。

原味達人的烹調祕笈

- 傳統烹調南瓜習慣切小塊再用水煮、清蒸或油炒。拌炒配料還會多加油及水，口感油膩不清爽，無法享受南瓜自然的原味。
- 南瓜的原味烹調時間比傳統作法快又簡單。以空間剛好的鍋具，不加鹽、不加油，讓南瓜利用少量水所產生的蒸汽煮的剛好熟透，大塊烹煮時更可以吃到南瓜的香Q原味並保留較多甜分。
- 配料用少油拌炒，味道很香，鋪在南瓜上，整體口感清爽美味，是一道可以吃到比傳統美味更好吃的健康料理。

Ingredients:

600g small sized pumpkin (orange skin), 150g pork picnic shoulder (0.5 cm shreds), 40g dried shrimps, 20g dried mushrooms (0.5 cm shreds), 15g garlic (minced), 2 tsp canola oil, dash scallion and fried shallots

Marinade:

2 tsp mushroom-flavored soy sauce, 1 tsp oyster sauce, 1/2 tsp sesame oil, 1/2 tsp white pepper, 1 tsp mixed juice of scallion, ginger, garlic and wine

Method:

1. Wash pumpkin, discard seeds, cut vertically into shape of a boat, 7 cm wide; marinate pork for 10 mins; place pumpkin skin-side down in a 2-quart pan, add 100 c.c. filtered water; bring to a boil on medium heat till much steam appears, turn to low heat and simmer for 6~8 mins till done.

2. While cooking pumpkin, preheat an 8-inch skillet till water droplets "dance slowly"; add garlic, 1 tsp oil, stir-fry on low heat till golden yellow, remove; add shrimps, mushrooms and fragrant garlic, cover, pan-fry each side for 1 min on medium low heat, turn to low heat, remove.

3. Add 1 tsp oil, spread pork evenly, cover; pan-fry for 1 min on medium fire; flip, top with ingredients in step 2, cover; pan-fry for 1/2 min on medium fire, give a quick stir; pour over the pumpkin, garnish with scallion and shallots.

Culinary Tips:

- Traditionally, pumpkin is cut into smaller pieces and parboiled, steam or stir-fried with oil, and then fried together with other ingredients with more oil and water, draining food of its natural goodness.
- Using minimum water to cook the pumpkin in larger pieces and perfect timing, without salt and oil in the right size pot, the valuable nutrients and natural sweetness is retained.
- Using minimum grease to fry the other ingredients separately makes this healthy dish very palatable; as a whole, the flavors are blended in a refreshing way.

全麥無酵鬆餅

15分鐘 / 4人份 / 1人份160卡

Healthy Whole Wheat Pancakes

15 mins / 4 servings / 160 cal per serving

材料：

雞蛋2顆（120克）、特調綜合鬆餅粉60克、碎核桃及葡萄乾適量、芥花油少許

作法：

1. 雞蛋打散放入容器中，以打蛋器由中速轉高速打至硬性發泡（蛋液往下滴呈鳥嘴型狀），再加入特調綜合鬆餅粉，以最慢速攪拌均勻（約10秒），用橡皮刀拌勻成鬆餅糊。

2. 利用打蛋時，取8吋平底煎鍋，加蓋，預熱至水珠慢跑，再把全部的鍋面及鍋邊均勻塗抹少許油（以不看到油紋為原則）。

3. 將碎核桃均勻撒在鍋底，再自鍋子中心點，均勻倒入雞蛋麵糊至鍋邊，加蓋，以小火烤約5分鐘，開蓋後，再放入適量葡萄乾，再加蓋繼續烤1～2分鐘至剛好熟，取出，用鋸齒刀切成6小塊即成。

原味達人的烹調祕笈

- 傳統製作鬆餅習慣用烤箱，水分會被烤箱燈管熱度吸收掉，口感較乾又易失去原味，必須使用較多油、水及發粉調整鬆軟度及濕度。

- 選對適當鍋具及完美的時間掌控，避開傳統烤法，完全不加油及發粉，貼在鍋底烤，可逼掉蛋腥味並利用雞蛋本身的水氣所產生的烤氣，保留蛋的溼度並產生金黃柔軟的鬆餅底，讓鬆餅吃起來綿密香Q，入口即化，還有濃郁的蛋香味。

- 特調綜合鬆餅粉的正確比例為全麥麵粉：低筋麵粉：細糖粉（或細楓糖粉）以1：1：2。如低筋麵粉2杯、細糖粉（或楓糖粉）4杯放入細濾網中，加入少許鹽一起過篩，再加入全麥麵粉2杯混合均勻即成。雞蛋的重量與鬆餅粉的完美比例是2：1。

Ingredients:

120g/2 eggs, 60g homemade pancake mix, dash walnuts (chopped), dash raisins, dash canola oil

Method:

1. Beat whole eggs at medium high speed until mixture is stiff, fluffy and bubbles becomes fine; add in homemade pancake mix, beat at lowest speed for 10 seconds, mix batter well manually with spatula.

2. While whisking eggs, preheat an 8-inch skillet with lid on till water droplets "dance slowly"; grease slightly the bottom and side of the pan, the grease streaks should not be visible.

3. Spread walnuts evenly on bottom of skillet, pour batter in the center of the skillet until it spreads to the edge, cover immediately; pan-grille on low heat for 5 mins, uncover, sprinkle raisins; grille for another 1~2 mins till done; cut with a jagged knife into 6 wedges and serve.

Culinary Tips:

· Traditionally, pancakes are baked in an over-spacious oven and the heating element tends to dry out the moisture, causing it difficult to swallow; also the natural flavor is greatly reduced; therefore, oil and self-rising flour is needed to adjust the texture and moisture of the pancake.

· Using the right size pan and with perfect timing, not only retains the natural moisture and flavor, it also preserves the rich, appealing melt-in-your-mouth tender texture; pan-grilling on the pan, covered, eliminates the unpleasant scent of egg, and allows the pancake to be baked in its own natural moisture and gives a golden brown and tender base.

· The homemade pancake mix is prepared with the volume ratio of whole wheat flour (1): low gluten flour (1): icing or fine maple sugar (2), (for example, 2 cups wheat flour, 2 cups low gluten flour, 4 cups sugar), sift the low gluten flour and sugar with dash salt, then mix well with whole wheat flour. The proportion weight of egg to pancake mix is 2:1.

附表1：一週家庭菜單規劃表 ▌ ATTACHMENT 1：WEEKLY FAMILY MENU PLAN

	星期一/Monday	星期二/Tuesday	星期三/Wednesday
早餐 Breakfast			
午餐 Lunch			
晚餐 Dinner			

（本表表格可重覆影印使用，或是您學會開立菜單之後，也可以把本表影印指導周遭的親朋好友，一起分享調整均衡的飲食，吃出健康的喜悅。）

| 星期四/Thursday | 星期五/Friday | 星期六/Saturday | 星期日/Sunday |

This chart can be photocopied and reused. After you have mastered creating weekly menus for your family, you can pass on the chart to your relatives and friends sharing with them to readjust their meals to a well-balanced nutritious diet and the joy of eating healthily.

附表2：我的家庭一週採買清單 | ATTACHMENT 2：WEEKLY GROCERY-SHOPPING LIST

供應商及電話 Grocery/Vendor Particulars	星期一 Monday	星期二 Tuesday	星期三 Wednesday	星期四 Thursday
豬肉 Pork				
牛肉 Beef				
羊肉 Lamb				
雞肉 Chicken				
魚肉 Fish				
水產 Seafood				
菇類 Mushrooms				
豆製品 Soy Products				
麵包 Bread				
蔬菜 Vegetables				
水果 Fruits				
雜貨 Grocery Store				
超市 Supermarket				
其他/Others				

（本表可重覆影印使用，或是您學會開立菜單之後，也可以把本表影印指導周遭的親朋好友，一起分享調整均衡的飲食，吃出健康的喜悅。）

星期五 Friday	星期六 Saturday	星期日 Sunday	本週總量 Total for this week	上週存貨 Previous balance	本週應購量 Purchase this week

This chart can be photocopied and reused. After you have mastered creating weekly menus for your family, you can pass on the chart to your relatives and friends sharing with them to readjust their meals to a well-balanced nutritious diet and the joy of eating healthily.

Acknowledgments

The process of writing this book was full of surprises (a filming on a Typhoon day till dawn almost ended in a fire), hardships (a bilingual book requires more than double the time and human resources), joy (seeing the final product), and pain (the hard labor of matching the two languages). Every step of the details from planning, writing, editing, photographing, translating to publishing require many people to join in at very specific stages to contribute their talent and energy in order to complete this cookbook.

We want to especially acknowledge the following people and organizations for their help and support:

Hanks for helping us with editing during our busiest time with filming; Our Chinese-to-English translation team: Roni, Andy, Grace, Aaron, and Kevin; May-Yuin, and Rei-Ling for editing the layout of the book; The editor, Hsiao-Ling and the marketing team, Ya-Wen and her colleagues who have always given their best for the promotion of our cookbook; The filming crew who stopped at nothing – even the Typhoon, storm, and rain; Margaret and the KC kitchen crew who assist Kevin in the photo sessions created the final products; KC Kitchen for reserving the location for photo sessions and providing cooking tools and utensils; Hugh Moeser and Karen Huang from the Canadian Trade Office in Taipei for sponsoring our trip to Canada and introducing us to Canadian producers; Valerie Chiasson, International Market Development Officer of the Agriculture and Agri-Food Canada, Markets and Trade, for showing us the characteristics and culture of Canadian food who is also our amazing tour guide on our tour across Canada; Canada Beef Export Federation for providing quality Canadian beef; The Taiwan Council of Agriculture (COA) for providing information on agriculture products; Dior Makeup Specialist for their help with the cover photo.

Finally, we want to give all the glory to the Lord our God who loves us. It was He who first gave us this healthy mission and then the strength to persevere. He sent His angels to help us put together our ideas and write a cookbook that can change the way people cook. We hope that through this cookbook we can pass on our blessings to all the families in their pursuit of healthy cuisine and enjoy the Natural and Heavenly Fare.

後記

　　製作這本書的過程，充滿了驚奇（拍攝時期，遇上颱風夜，拍到凌晨還差點引起火災），歷盡了艱辛（中英雙胞胎，時間、人力等多了一倍以上的付出），有歡笑（成品出爐時），有痛苦（中英校對時幾乎要難產），從企劃成型、文字整理、成品拍攝、情境捕捉、中翻英、中英校對編譯等，所有的環節都因為有許多人的投入與付出，才能一一完成。

　　我們要特別感謝下列人士暨機構單位給予這本書的協助：

- 在我們電視錄影最忙碌時，協助文字總整理的玉春（Hanks）。
- 協助將中文翻譯成中文的英譯團隊Roni、Andy、Grace & Kevin。
- 協助編排整本書的美雲及美編瑞玲。
- 協助編輯企劃的小鈴、極力推廣本書的雅雯及行銷企劃團隊。
- 在颱風天，風雨無阻拍攝到底的攝影團隊。
- 協助Kevin製作成品的Margaret及KC健康廚房工作團隊。
- 完全不營業，提供攝影場地及烹飪道具的KC健康廚房。
- 協助KC認識加拿大食材供應商的「加拿大駐台貿易辦事處」副處長及資深經理黃彗怡。
- 協助Kevin & Claire認識加拿大當地食材特性、飲食文化交流的「加拿大農業及農產食品部國際市場局」，及一路陪到底的超級饗導夏文麗專員。
- 提供加拿大優質牛肉的「加拿大牛肉出口協會」臺灣辦事處。
- 提供優良農場品諮詢的臺灣農委會。
- 協助封面人物彩妝設計的迪奧彩妝師。

　　最後要將一切榮耀歸於愛我們的上帝，因為　祂所賦予KC使命去製作能影響人改變烹調習慣的食譜，給KC力量堅持到最後，並帶來許多天使共襄盛舉，盼望透過這本書祝福所有追求健康烹調的家庭。

Family 健康飲食 12Y

KC健康廚房 從零開始學做菜！向食物借油╳借水的健康美味烹調法

作　　　者／Kevin & Claire
選　　　書／林小鈴
主　　　編／陳玉春

行銷企畫／林明慧
行銷副理／王維君
業務經理／羅越華
總 編 輯／林小鈴
發 行 人／何飛鵬
出　　　版／原水文化
　　　　　台北市民生東路二段141號8樓
　　　　　電話：（02）2500-7008　　傳真：（02）2502-7676
　　　　　E-mail：H2O@cite.com.tw　部落格：http://citeh2o.pixnet.net/blog/
發　　　行／英屬蓋曼群島商家庭傳媒股份有限公司城邦分公司
　　　　　台北市中山區民生東路二段141號11樓
　　　　　書虫客服服務專線：02-25007718；25007719
　　　　　24小時傳真專線：02-25001990；25001991
　　　　　服務時間：週一至週五上午09:30～12:00；下午13:30～17:00
　　　　　讀者服務信箱：service@readingclub.com.tw
劃撥帳號／19863813；戶名：書虫股份有限公司
香港發行／城邦（香港）出版集團有限公司
　　　　　香港灣仔駱克道193號東超商業中心1樓
　　　　　電話：(852)2508-6231　　傳真：(852)2578-9337
　　　　　電郵：hkcite@biznetvigator.com
馬新發行／城邦（馬新）出版集團
　　　　　41, JalanRadinAnum, Bandar Baru Sri Petaling,
　　　　　57000 Kuala Lumpur, Malaysia.
　　　　　電話：(603) 90578822 傳真：(603) 90576622
　　　　　電郵：cite@cite.com.my

城邦讀書花園
www.cite.com.tw

熱量分析／徐于淑
英文編譯／Andy・Roni・Kevin・Grac
美術設計／許瑞玲
特約攝影／子宇影像工作室・宋和憬
製版印刷／科億資訊科技有限公司
初　　　版／2008年4月25日
二版一刷／2011年1月24日
二版三刷／2014年7月17日
三版初刷／2018年12月27日
定　　　價／450元
ISBN 978-986-6379-46-8
EAN 471-770-290-531-6

國家圖書館出版品預行編目(CIP)資料

KC健康廚房 從零開始學做菜！向食物借油
╳借水的健康美味烹調法 / Kevin, Claire合
著. -- 修訂初版. -- 臺北市：原水文化出版：
家庭傳媒城邦分公司發行, 2018.12
　面；　公分. -- (Family健康飲食；12Y)
ISBN 978-986-6379-46-8(平裝)

1.食譜 2.烹飪
427.1　　　　　　　　　　　　100000001